Sea Floor Stories

By
D. Connelly

Illustrations
Lyn Reed

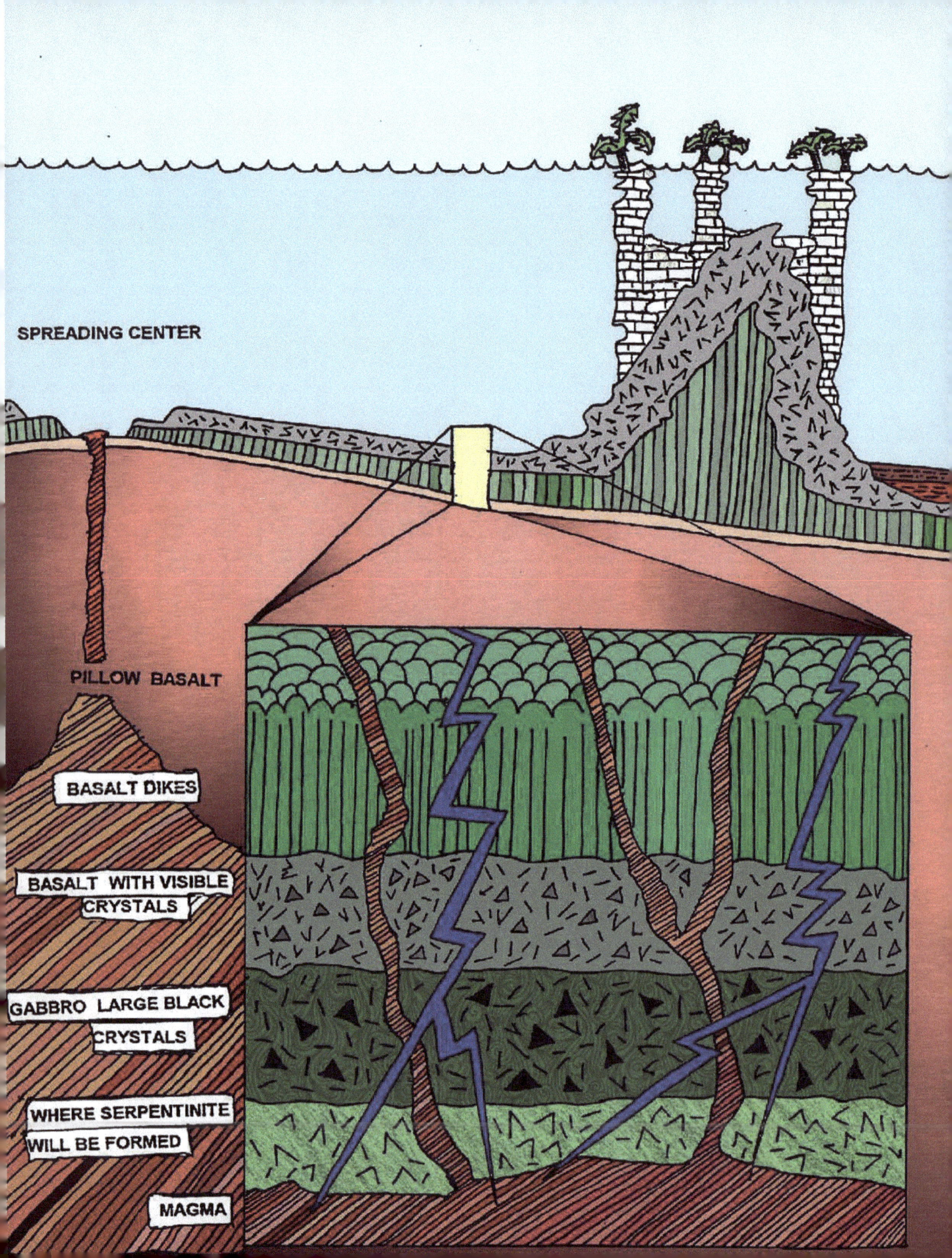
SPREADING CENTER
PILLOW BASALT
BASALT DIKES
BASALT WITH VISIBLE CRYSTALS
GABBRO LARGE BLACK CRYSTALS
WHERE SERPENTINITE WILL BE FORMED
MAGMA

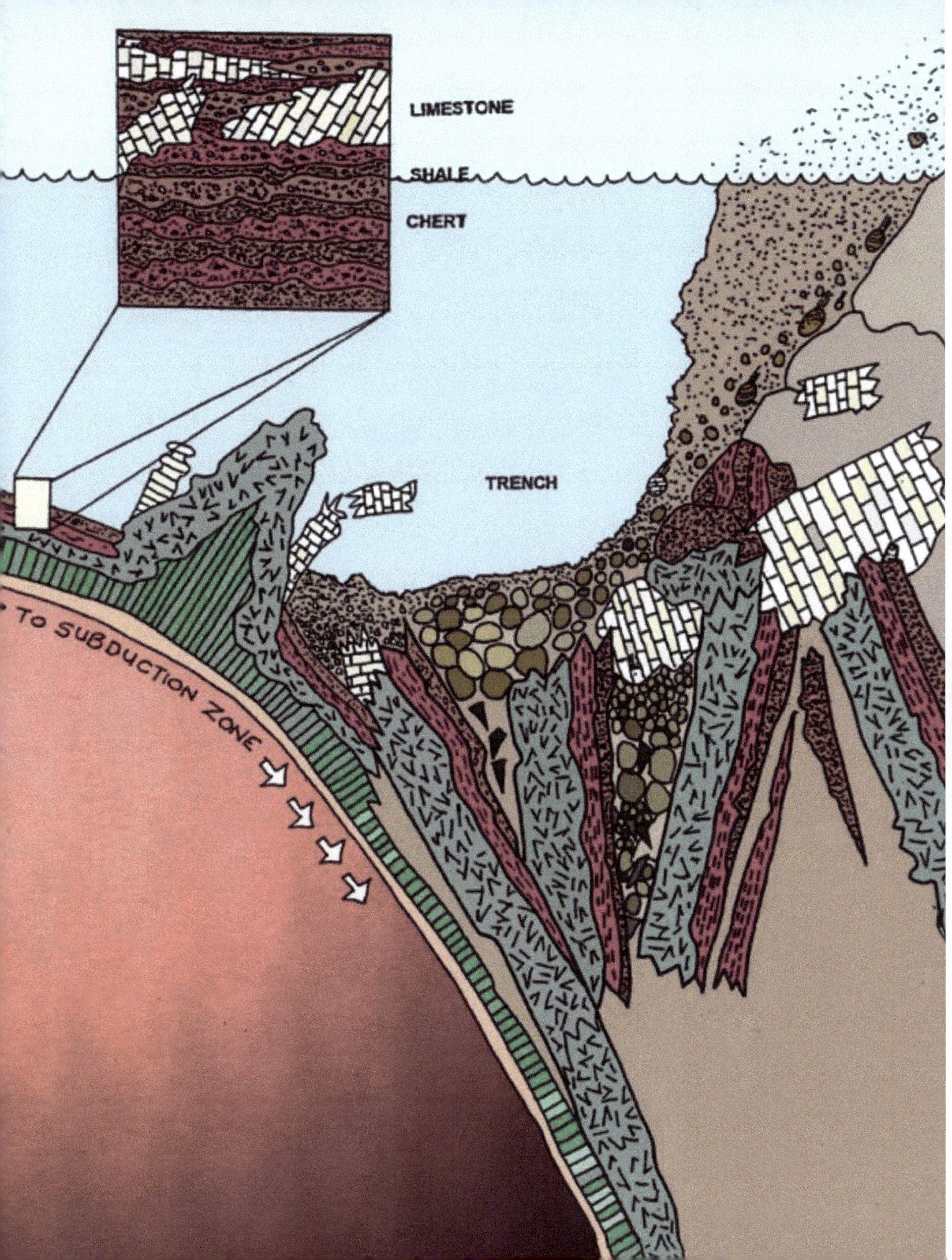
LIMESTONE
SHALE
CHERT
TRENCH
TO SUBDUCTION ZONE

This book is a work of fiction (or nonfiction). Names, characters, places, and incidents are the product of the author's imagination (or are used fictitiously). Any resemblance to actual events, locales, or persons, living or dead, is entirely coincidental.

Copyright number: 1-15025178191

ISBN:
Paperback: 9798349548819
Hardcover: 9798349556753

Library of Congress Control Number: 2025-000789

To the
Originator
of
Plate Tectonics

Table of Contents

Introduction

The mountains and hills that make up the coast range of California
contain some unusual rocks.
These rocks remained mysterious for a long time
because many of them came from the deep,
dark, secret places of the ocean floor.
This is the story of seafloor rocks and
how they ended up in the hills and mountains
all along the coastline.
The hills were caused by continuous pressure and collisions
between the ocean floor and the edge of the continent.
If we could have seen the rocks forming on the ocean floor,
we would have recognized them when we found them in the hills.

There was a time not so very long ago when no one could look at the rocks on the ocean floor. The floor of the ocean can be two or three miles underwater. Most of the ocean floor is so deep that no light ever reaches the bottom. That much water is heavy enough to crush a submarine. The ocean floor was dark, cold, dangerous and unknown. We figured out how to get to the bottom of the ocean about the same time as we figured out how to get to the Moon.

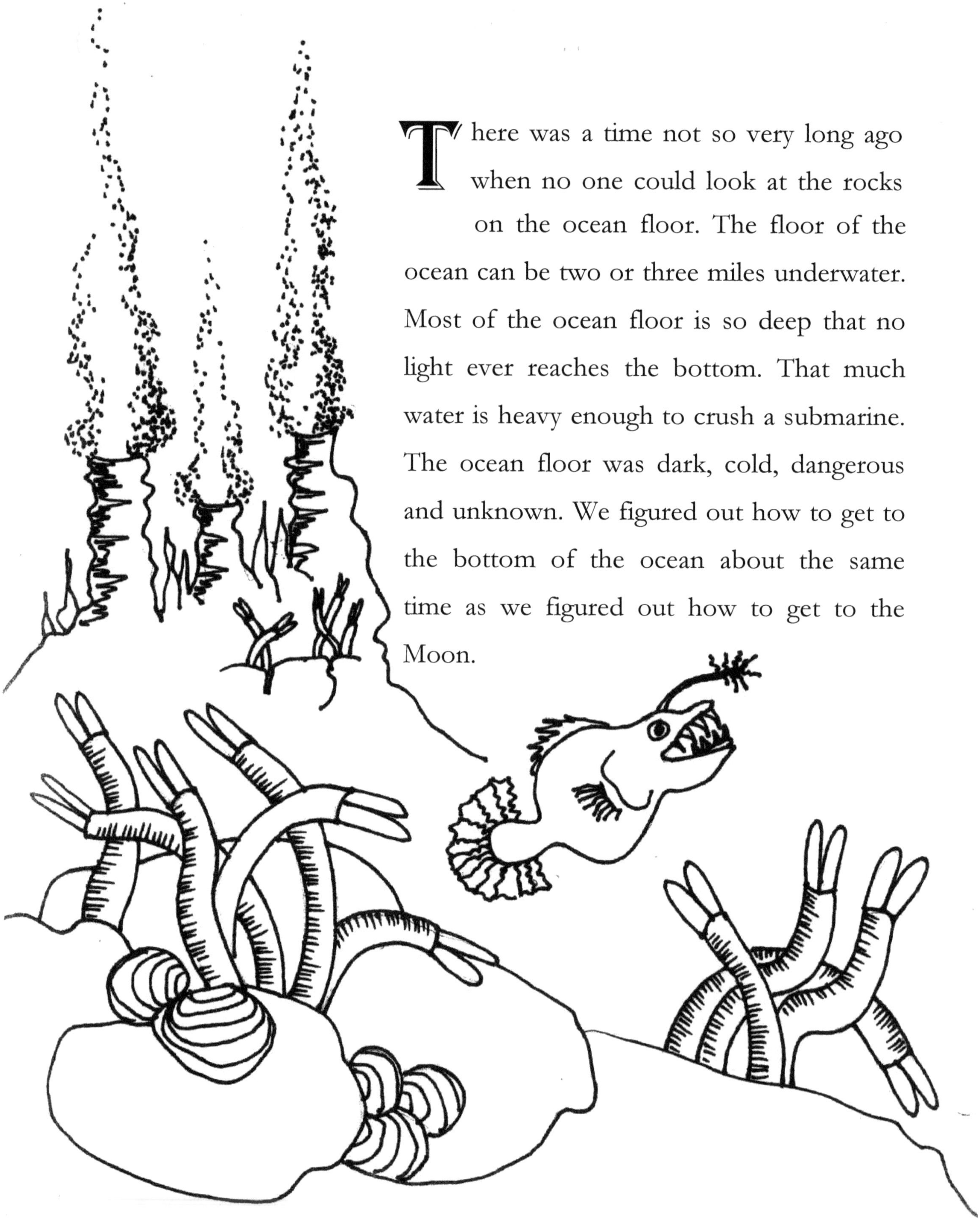

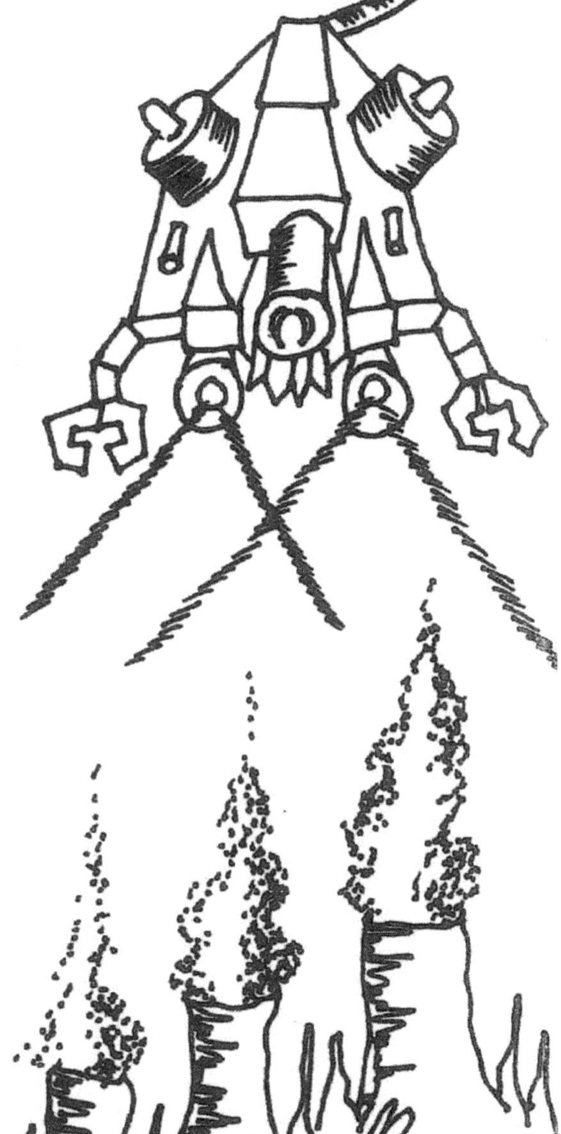

During the Second World War, ships gathered information about the ocean floor while they were out looking for submarines. Scientists were too busy with the war to figure out what the new information meant. After the war, in the 1950s and 60's, scientists had time and money for ocean research. They found ways to go into the ice-cold darkness and see volcanoes erupt. They built equipment to get to the ocean floor and back without being crushed. They figured out how to drill holes and bring back samples of the ocean floor.

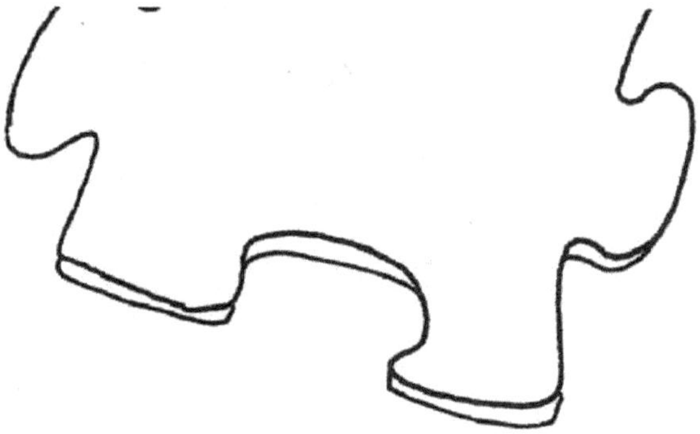

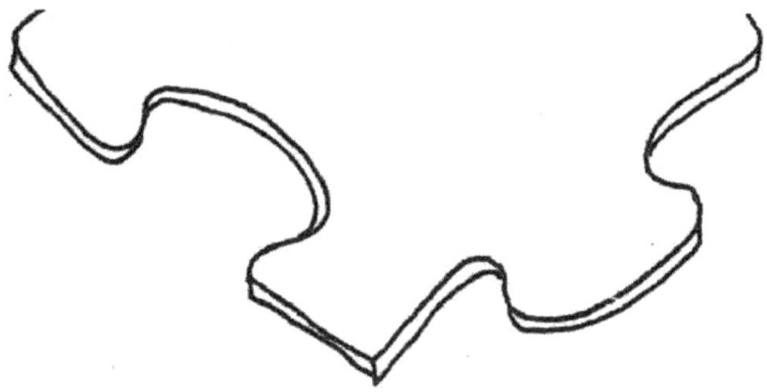

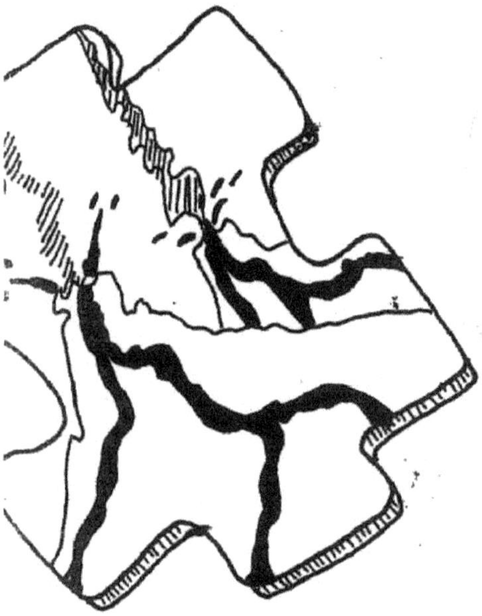

Each new bit of information was like a piece of a puzzle that tells the story of drifting continents and an ocean floor that grows and moves. Some of the pieces are still missing. Now, we know enough to understand some of the secrets of the rocks in the hills. The rocks tell us of collisions between the ocean floor and the land. Rocks record stories of eruptions of molten lava on the ocean floor or long, lazy years in warm tropical waters. The rocks have had marvelous adventures. You can read the stories in the rocks if you look carefully.

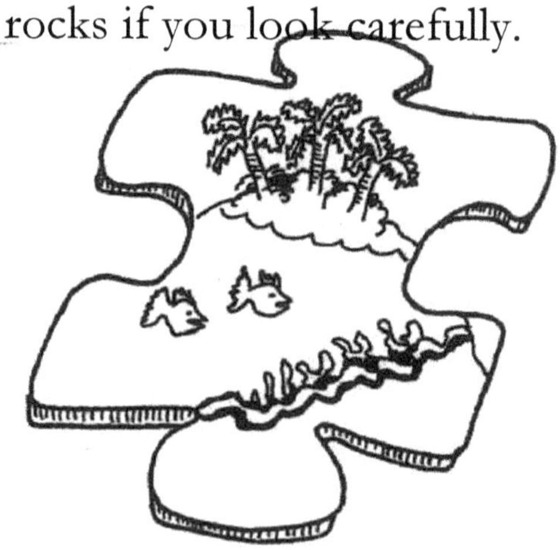

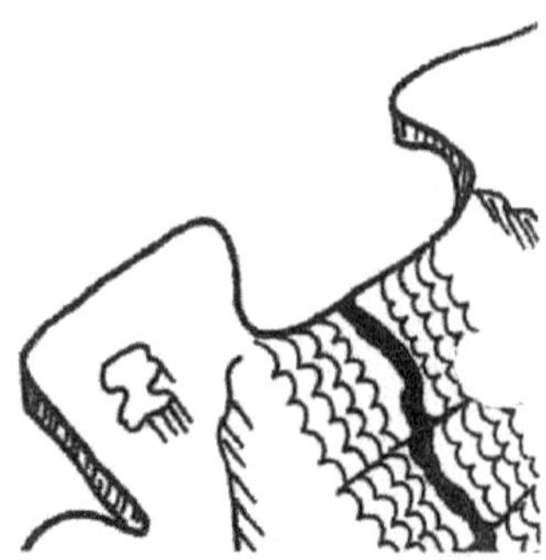

You need to notice a few things about rocks before they will tell their secrets. It helps to wash them off so you can see the details and color. It also helps to look inside. Differences are more obvious on the inside than on the weathered outside. If you break the rocks open, you can get a better view.

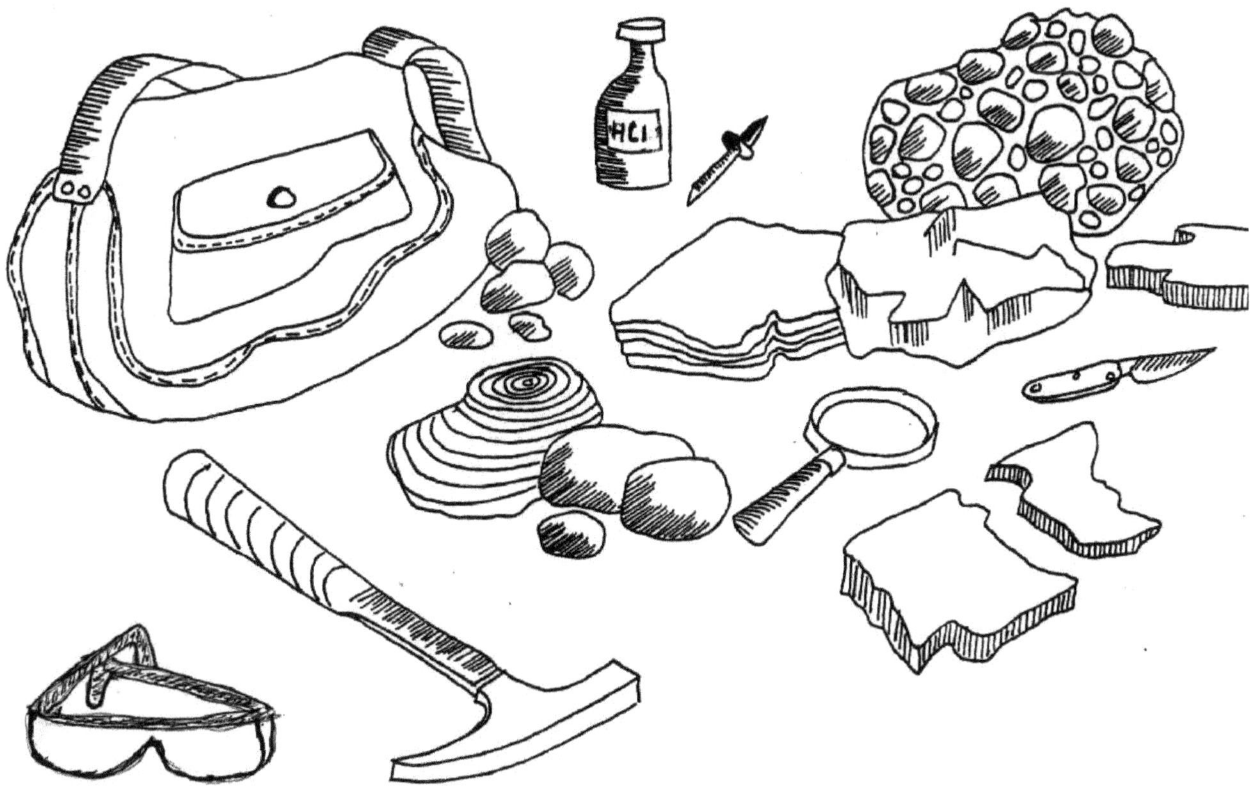

There are three main types of rock: igneous, sedimentary and metamorphic. Igneous rocks are made from melted rock or magma. Hot magma looks red, like fire. Once the red-hot magma pushes out onto the ocean floor or the land, it is known as lava. Rocks that harden from this hot liquid are called igneous rocks. The word " igneous" comes from a Latin word that means "fire".

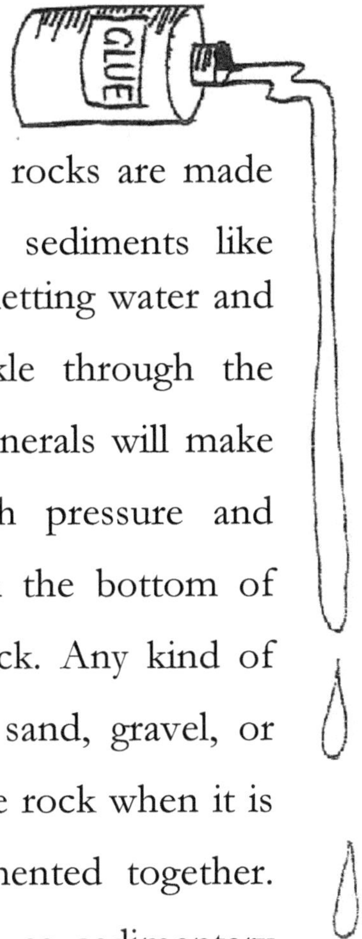

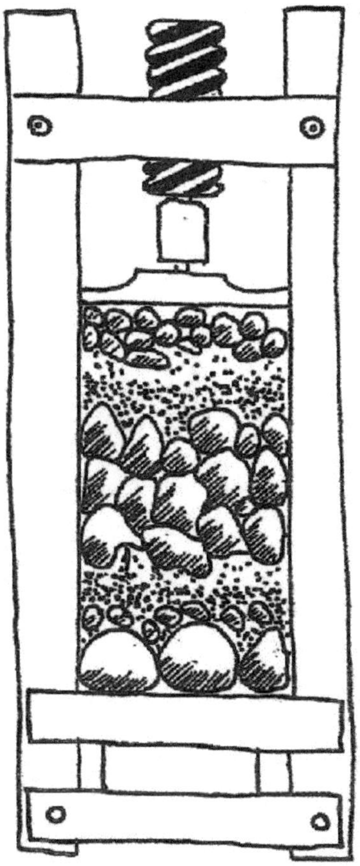

Another way that rocks are made is by piling up sediments like sand and clay and then letting water and dissolved minerals trickle through the layers. The dissolved minerals will make a natural cement. With pressure and cementing, sediment on the bottom of the pile will become rock. Any kind of sediment like clay, silt, sand, gravel, or volcanic ash will become rock when it is pressed down and cemented together. These rocks are known as sedimentary rocks.

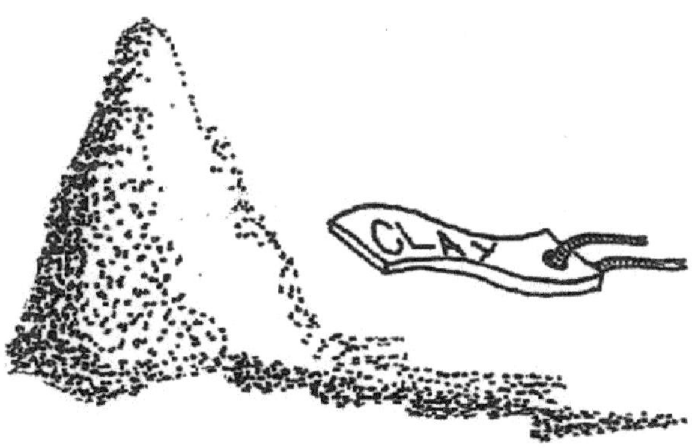

The third way that rocks are made is by being *changed* from an igneous rock or a sedimentary rock *by heat and pressure.* Because the rocks change from one type to another, like a caterpillar to a butterfly, they are known as metamorphic rocks. The change is caused by the heat and pressure deep within the earth. Rocks found in the hills formed where the ocean floor and continents collided include all three types of rock: igneous, sedimentary and metamorphic.

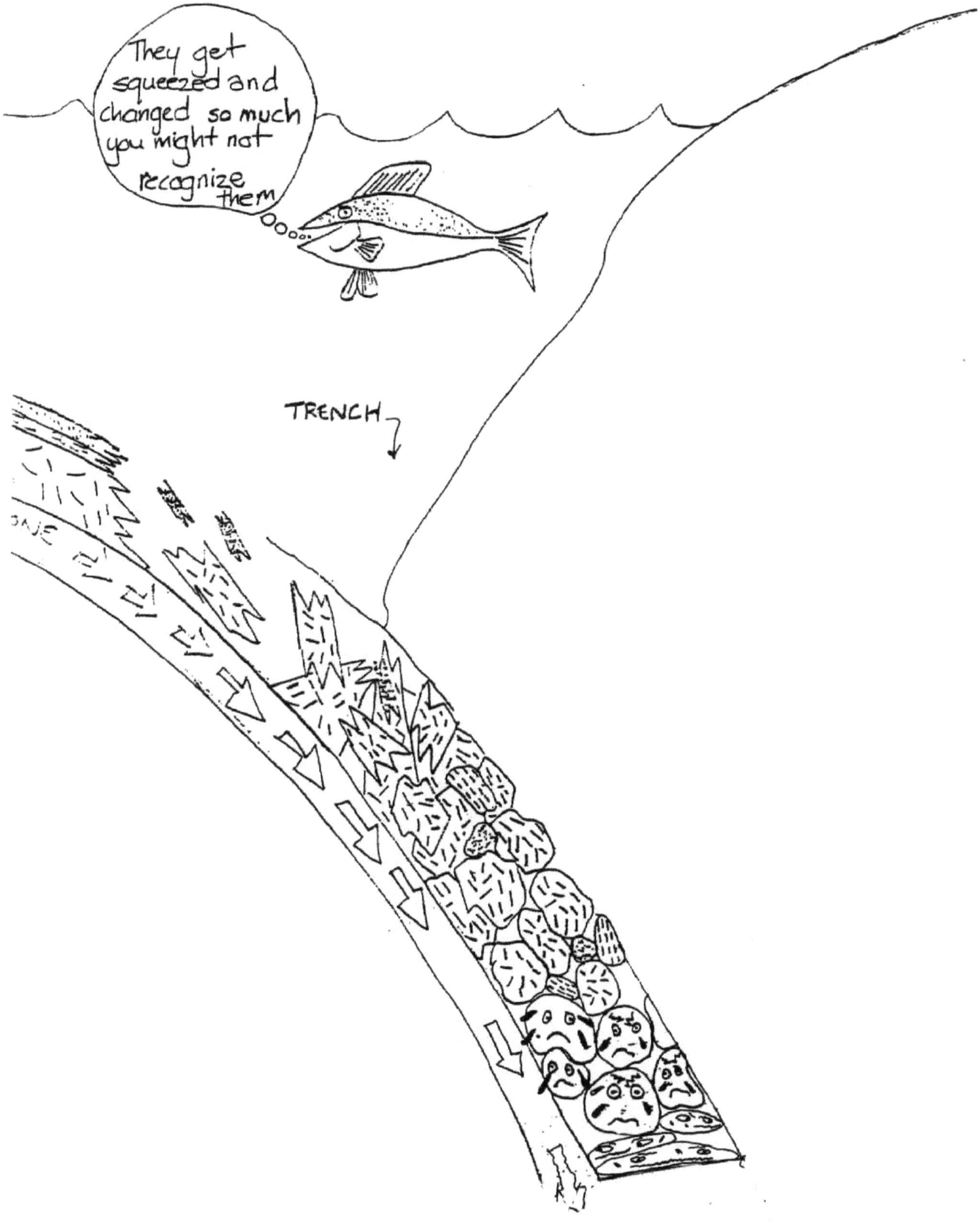
They get squeezed and changed so much you might not recognize them
TRENCH

GREENSTONE (A METAMORPHIC ROCK)

Greenstone, this rock has a marvelous adventure locked inside it. The story begins with an eruption on the ocean floor. Beneath thousands of feet of ocean water, moltenrock is pushed up and out onto the surface. The older solid ocean floor moves away slowly. This area where the new ocean floor is made is known as the spreading center. The solid ocean floor moves on the warm, flexible mantle layer beneath it. Greenstone was once a piece of the new ocean floor.

The rock slowly moved sideways, away from the spreading center. The whole ocean floor, with all of its rocks, is taking a long, slow ride. Rocks have plenty of time. They tell very old stories. The scientific name of the rock on the new ocean floor is basalt. If you could reach down to the bottom of the ocean and pick up a new rock, you would see that it is black. Somewhere between the spreading center where it is formed a nd the hills and mountains where it is found, the minerals in this rock will react with heat and chemicals in seawater to form new minerals that are green.

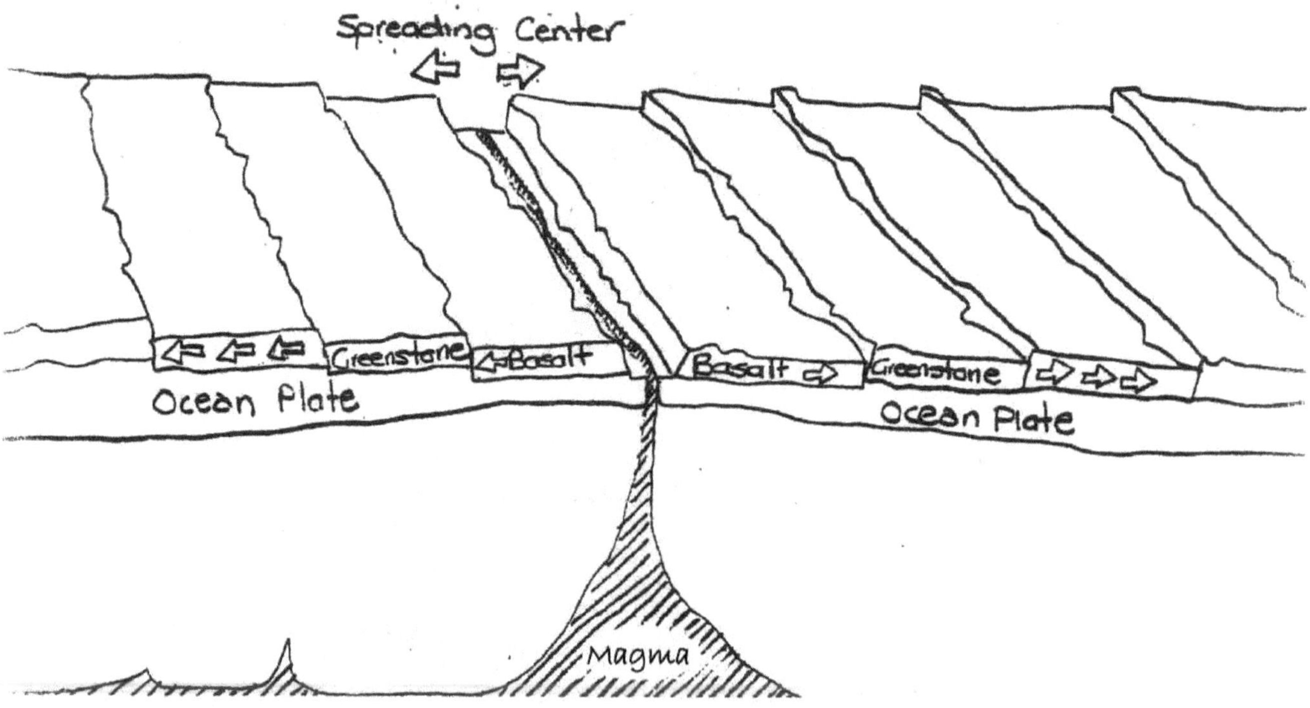

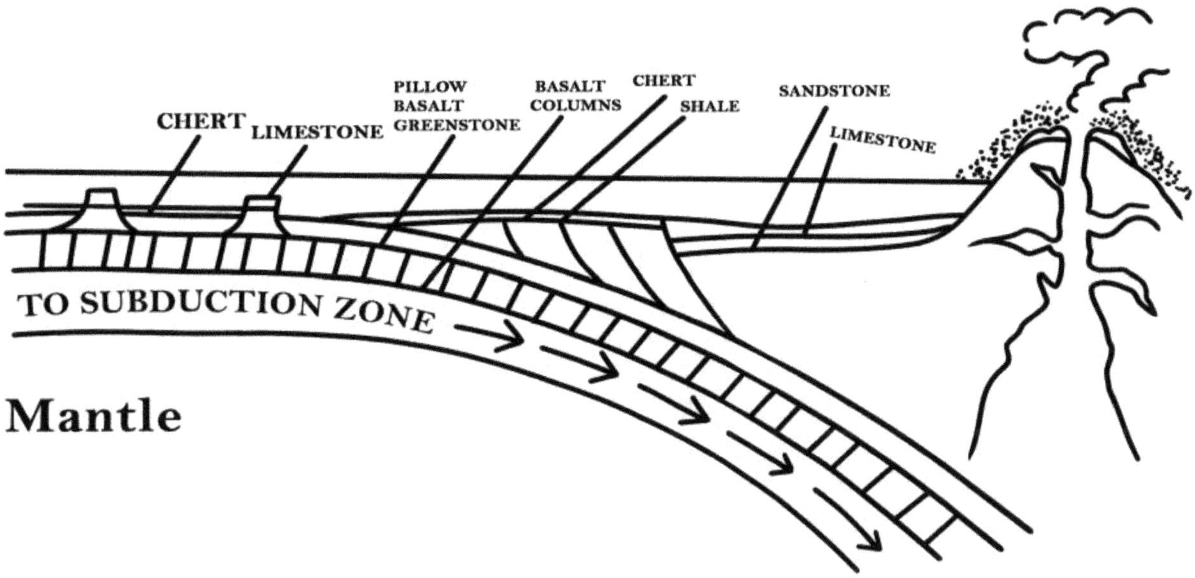

The ocean floor reaches all the way to the edge of the continent. The big, solid, slow-moving block of ocean floor pushes against the continent continuously. When we feel something bend or break in all this pushing, we call it an earthquake. Here at the meeting of the ocean floor and the continent, the ocean floor ducks under the thicker, lighter (less dense) block of land that is the continent. This area where the ocean floor bends and sinks below the continent into the softer, hot mantle of the earth is known as the **subduction zone**. Tons and miles of rock are bent, crushed and piled up as the solid ocean slab sinks into the hot mantle of the earth. Here, minerals can be changed by heat and pressure. Some of the rocks get broken off and pushed up out of the way by rocks coming in behind them. That is what happened to the greenstone. The greenstone was scraped off as the ocean floor bent and sank below the continent.

On the surface, the land is pushed up into hills like a rug when you slide into it. All the movement below causes hills and valleys to form on the surface. Millions of years later, the greenstone may be part of a mountain. The rain will help cut deep valleys into the mountain. The steeper the mountain, the faster the rain and the rivers can carry pieces of it away. Rocks that were once buried very deep get closer to the surface. One day, the greenstone appears in plain sight. The rocks that were once part of the basalt of the ocean floor now appear on the land as greenstone. When you find them, they may be weathered a bit orange and look dirty on the outside. If they are wet from the rain or the water in a creek, the rocks will look greener.

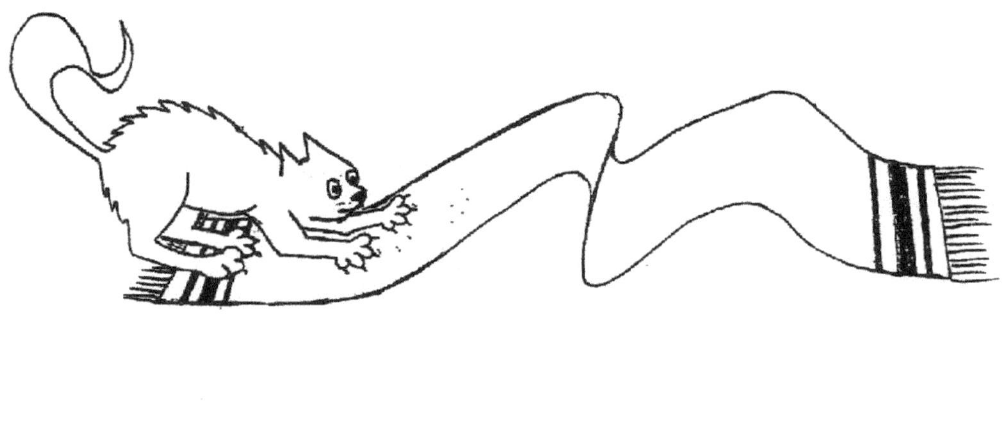

There is one member of this group of green rocks that is easy to recognize. This rock has light green spots. It might remind you of a spotted bird's egg. One way to get a speckled rock is from an eruption in shallow water. Holes from escaping gas will remain as the lava quickly cools to become solid rock. Some holes will fill with hard minerals that are in the ocean water. Underneath the ocean floor, where it is hotter, the rock can cool more slowly. Crystals have a chance to grow. This spotted green stone cooled slowly enough for the holes to fill. In some cases, you can even see the shape of the crystals. If you could see the rock when it was new, it would look mostly black with light crystals. When you find the rock on land, it might look rusty green with holes or light green spots.

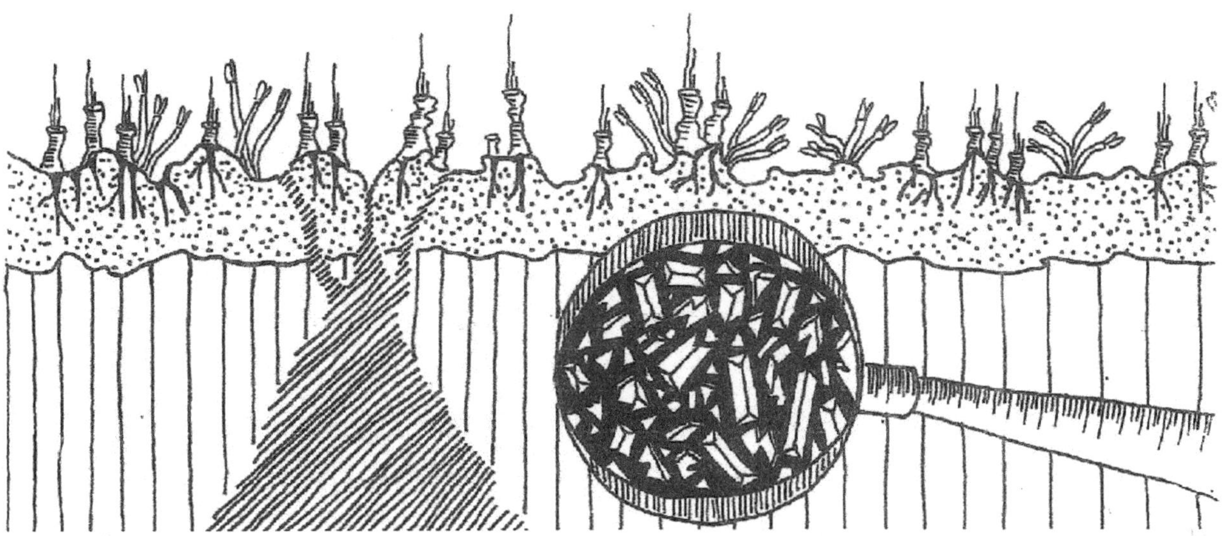

CHERT(A SEDIMENTARY ROCK)

Chert beds are one of the most colorful pieces of the puzzle made by Subduction zone rocks. These rocks are very hard and smooth. They are made from minerals

too small to feel or see. Many are about the color of red brick. Arrowheads made from chert last a long time because the rock is so hard. To really understand the story of chert, we must begin in the ocean with microscopic creatures called **plankton**. These tiny creatures live in the top layer of the ocean. A type of plankton called **radiolarian**s can make their skeletons from silicon dioxide (SiO_2), which is dissolved in the cold ocean water. You make your skeleton from calcium. Radiolarian skeletons and glass are made from the same material. Radiolarian skeletons are harder and last longer than yours. The microscopic skeletons of these plankton eventually sink to the ocean floor. The skeletons are deposited day after day, year after year.

The nooks and crannies of the ocean floor are filled with minerals and **silt** that drift down from the surface of the ocean. Silt is made of tiny pieces, smaller than sand, like the dust that the wind blows. Silt and skeletons from radiolarians make layers of rock all across the ocean floor. Chert and silt can pile up in layers for the whole lifetime of the ocean floor. Scientists estimate that chert could pile up one yard deep in the time it takes to add five or ten miles of new ocean floor. The little radiolarian skeletons found in the ocean are white.

Most of the chert is red. How can we explain that.

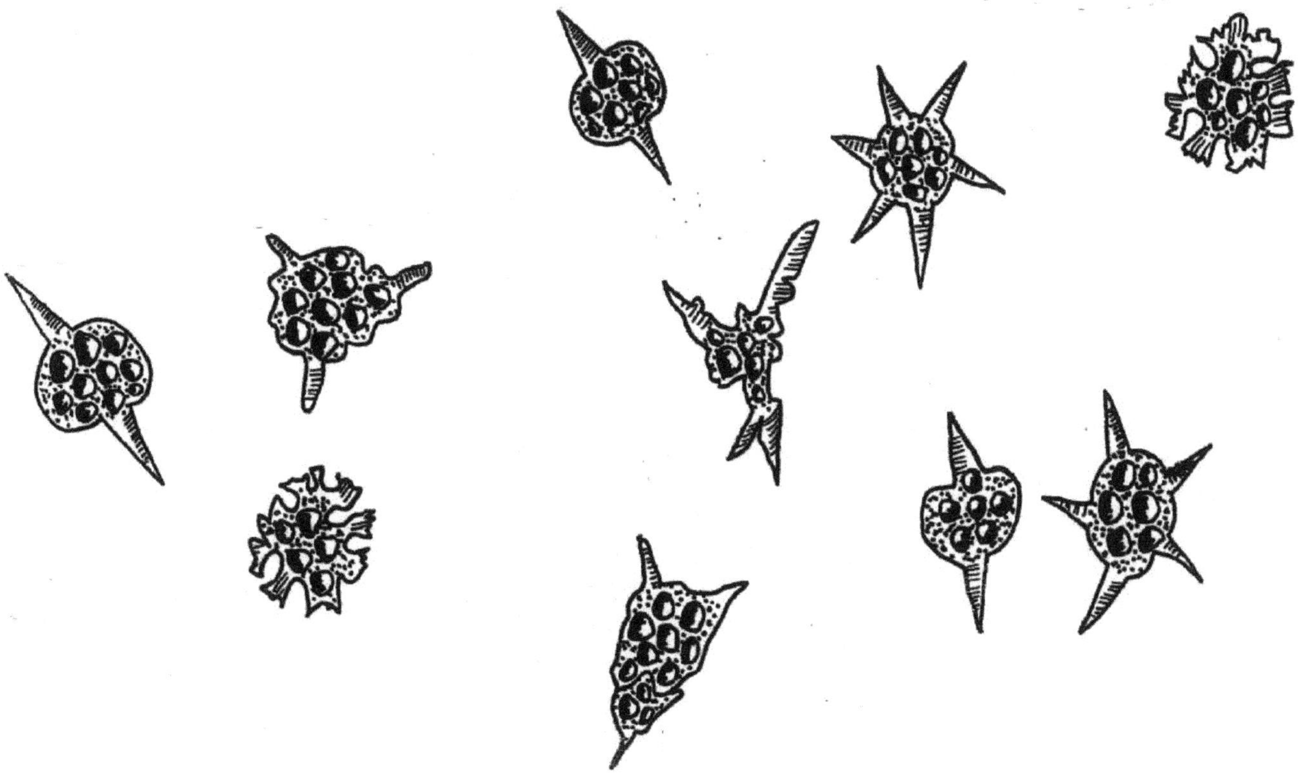

If iron and oxygen were found together, we would see that red color. Most of the rocks that make up the continents are made of elements that are light in weight and color, like silica and aluminum. If you could walk around on the new ocean floor, you would notice that the rocks were much darker than rocks from the land. That is because the ocean floor is mostly made of dark minerals that contain the element iron. Iron can dissolve in all that ocean water. Iron also comes from the dead things that sink to the ocean floor. Iron is in the dust that wind carries to the ocean from the continents. Very cold water can hold a lot of oxygen. The coldest water sinks to the bottom and brings oxygen with it. When iron and oxygen combine, you may see a reddish color, like the color of rusty nails. You find rust wherever iron, water and oxygen are found together. When silica combines with iron and oxygen, red chert can form.

New cooled ocean floor has a constant supply of things falling on it from above. Think what would happen under the kitchen table if nobody swept for a million years. The Ocean floor, with all the stuff that has fallen on it, is moving slowly away from the spreading center where the new ocean floor is made. It slowly moves toward the subduction zone, where the ocean floor is dragged under the continent. The ocean floor is carrying layers of chert and silt like frosting on a cake.

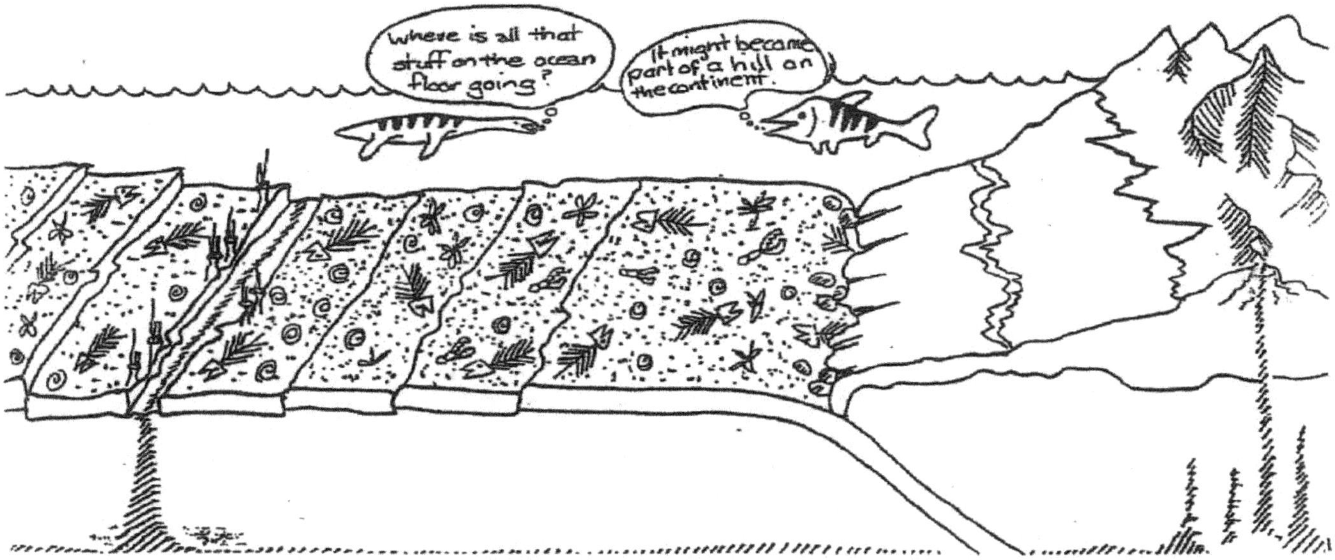

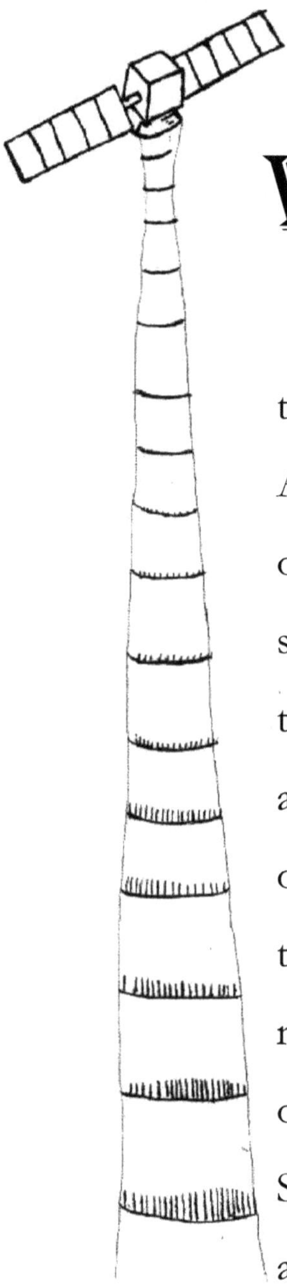

W What happens when the ocean floor pushes against the land? If we could take one picture every 10,000 years from the Moon and make a movie, the land would look like it was bending and folding. Along the edge of the continent where the collision occurs, hills are formed. During the ride toward the subduction zone, cold rocks from the ocean floor begin to move under the continent. The rocks may get warm and bend. Many rocks get scraped off and pushed up out of the way. Wherever the collisions occur, rocks that used to be on the ocean floor can now be found in the mountains and hills of the continent. The rivers flowing out of the mountains cut through and polish the rocks. Some of these rocks can be found in your backyard. Long ago, when the coastline was being folded and bent, some of the chert from the ocean floor was broken off and pushed up into a hill. Later, the rocks were uncovered by creeks cutting through the hills and by children digging in the backyard. Red chert is one of these rocks.

If you have found chert, you have also found a rock that spent a long time underwater. Some of these rocks were formed on the ocean floor when dinosaurs were rambling around on the land and swimming in the oceans. You can recognize chert by its hardness and smoothness. Chert is not always red. Chert is sometimes green or grey. The color of the chert depends on things like the amount of oxygen that is available and which other elements are around to mix with the silica when the rock is forming.

SHALE (A SEDIMENTARY ROCK)

To make shale, you begin with pieces that are too small to see. When you rub clay between your fingers, you cannot feel the very small pieces. When these pieces are pressed down and cemented together the rock is known as shale. Shale comes from pieces of sediment that are so tiny the wind can carry them all the way to the ocean. Volcanoes can blow clay-sized pieces so far up into the air that the dust may settle a thousand miles away. Some mineral dust will eventually find its way to the ocean floor. When rivers carry clay to the ocean, the tiny pieces float far out to sea and travel with the currents. These tiny pieces of sediment will settle to the bottom and become part of the **ooze** on the ocean floor. Your toes like to wiggle in ooze. A shallow, muddy ooze is a good place for clams and their friends to live. The clay minerals can stick together and form layers on top of the sea shells and plankton left in the ooze. As the layers become deeper, the muddy clay on the bottom turns into a rock called shale. The shale continues on its journey with the other rocks of the ocean floor back to the edge of the continent. The speed of this trip has been estimated to be about an inch a year on average, or about as fast as your fingernails grow. Eventually, this rock, with its collection of fossils, will find its way across the trench and up into the hills along the coast.

Look at all that red radialarian ooze and mineral dust accumulating on the ocean floor
And Other Plankton too
23

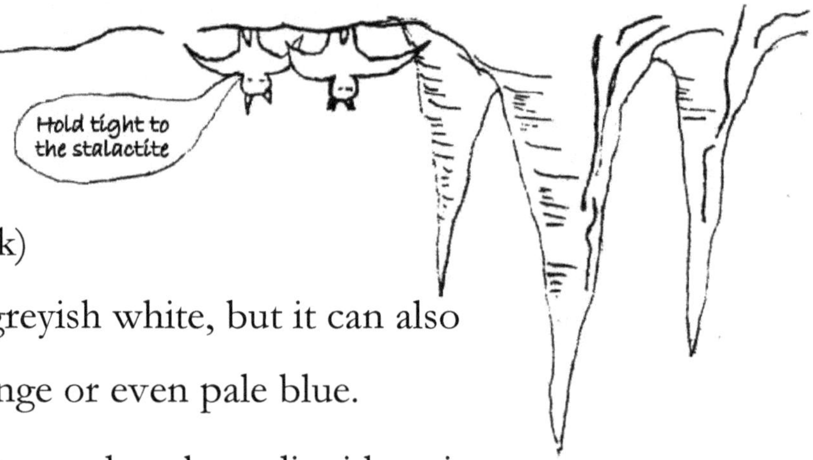

LIMESTONE

(a sedimentary rock)

This rock is most often greyish white, but it can also be black, brown, pale orange or even pale blue. Limestone can dissolve when water and carbon dioxide mix. When limestone dissolves underground, groundwater carries away the solution. The missing part gradually becomes a cave. Later, the water evaporates, and the dissolved limestone reappears somewhere else. It may appear as a stalagmite on the floor of a cave, or it may fill the spaces between grains of sand or gravel. The limestone will have provided cement to make a new rock. Limestone is not hard compared to chert or greenstone. They will both scratch limestone. If you put vinegar or some other acid on a rock and you see tiny bubbles, it is a good clue that you are looking at limestone. Almost every state in the USA has limestone caves.

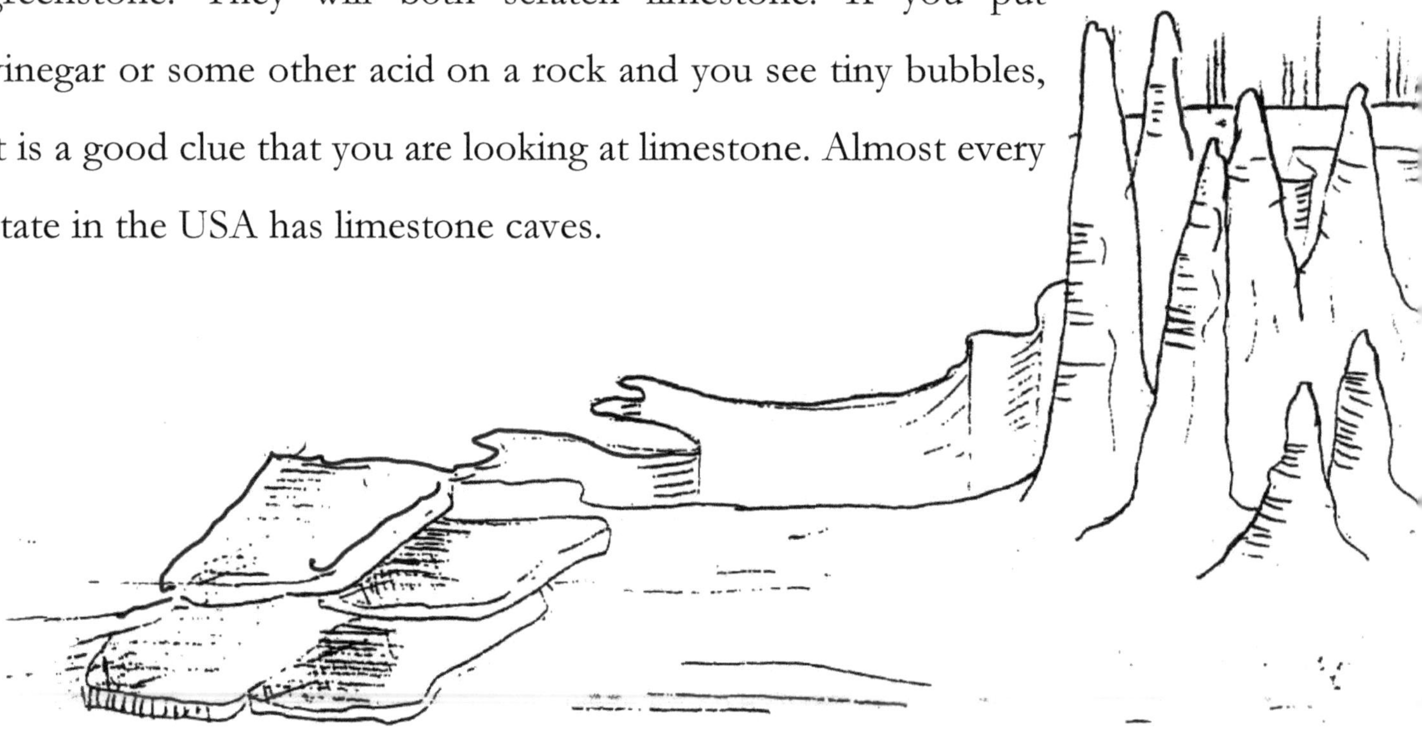

How did limestone get into the hills?

The story takes us back to the ocean. The limestone that we find in the hills today is made from the shells and skeletons of the creatures that lived in the warm surface waters. When the water is warm and not too deep, the skeletons of the creatures accumulate. Today, plankton skeletons pile up on the tops of undersea mountain ranges like snow. Skeletons pile up offshore in shallow water. Long ago, in places where the deep ocean was warmer, skeletons even piled up on the sea floor. Water can dissolve these skeletons if it is very cold or very deep. Our future limestone will either be buried by sediment so it is out of the reach of cold deep water, or it will form in shallower water close to shore.

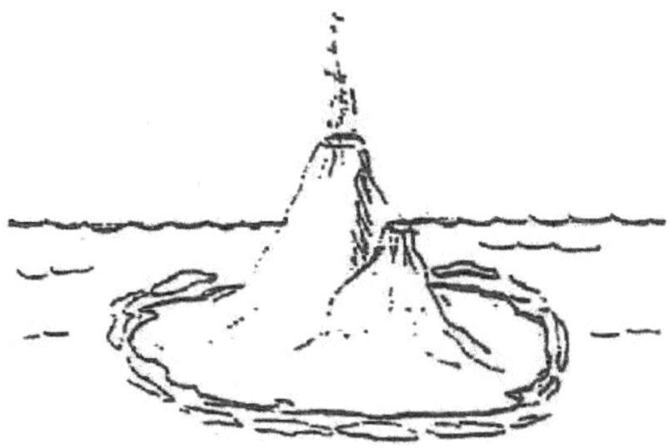

The coral anchors itself underwater to the sides of the volcano.

the volcano is weathered away.

The coral reef continues to grow

Today we find coral reefs growing in the in warm, shallow waters of the tropics. If the sea level changes, the creatures that make up the reef simply grow in a better location. The best place for the sea creatures to live is along the edges of the reef. There are lots of things to eat and good places to hide at the edge of the reef. The coral reef itself is made of the exoskeletons (outside skeleton)) of tiny animals.

Your skeleton is inside. You can feel your skeleton, but you can't see it. The skeletons of the coral animals are on the outside. If you go to a coral reef, you can see the exoskeletons. If you want to see the living coral, you need to look early in the morning or at night. The coral animals tuck inside their rough exoskeletons. Very few fish can eat them when they are hiding. The creatures that build the reef feed on bits of food that are carried by currents and tides. The tiny creatures grow and thrive at the edge of the reef.

As long as the temperature is right and the water is not too deep, the reef will grow. The reef grows like a continuous block of tiny apartments. Each little coral animal wants a room with a nice food supply within easy reach. New apartments are constantly added to the outside. Other creatures of the sea, like clams a nd snails, live in the reef. Their shells are made of the mineral calcite just like limestone. As the outside of the reef grows, the inside dies. The holes in the reef are filled in with the dissolved mineral, calcite. When the old reef becomes a solid block, we can call it limestone.

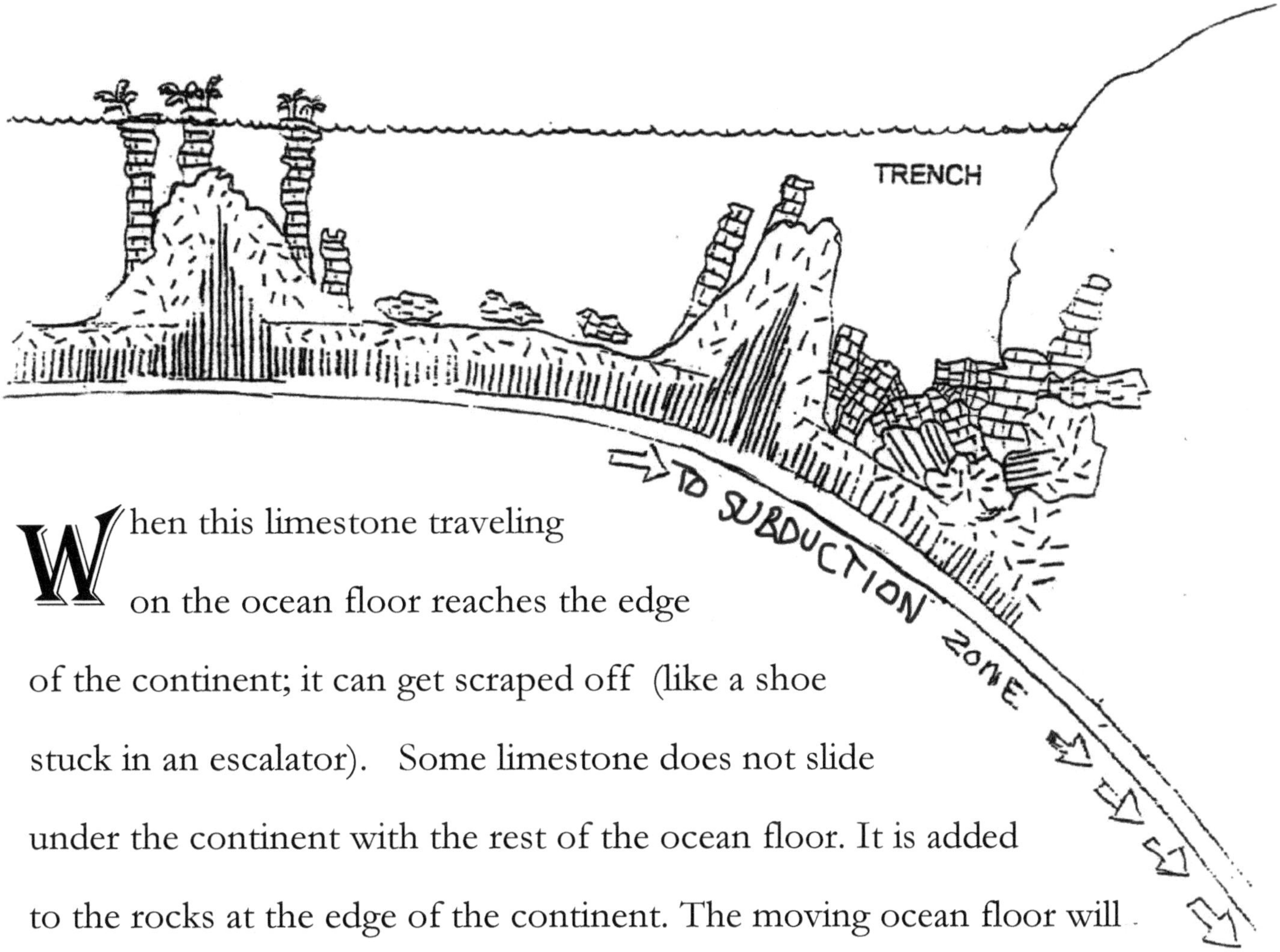

When this limestone traveling on the ocean floor reaches the edge of the continent; it can get scraped off (like a shoe stuck in an escalator). Some limestone does not slide under the continent with the rest of the ocean floor. It is added to the rocks at the edge of the continent. The moving ocean floor will eventually lift the limestone. The limestone will become part of the hills. Sometimes you can see the shape of the animal shells that made up the limestone. You can find pieces of limestone where the creeks have cut through the hills. You can find limestone in the creek beds when the rocks are too big for the creek to carry away. Some creeks are full of this rock. If you follow such a creek up into the hills, you may come to a quarry where they crush and heat limestone to make cement. You can find limestone wherever there once was a warm tropical sea.

CONGLOMERATE

Conglomerate is a sedimentary rock made up of large, rounded pieces of older rocks. The rock pieces may be almost any color, size or type. It is a conglomeration of other rocks all stuck together. It can form wherever rounded rocks have a chance to pile up. Rivers and waves are very good at making piles of rounded rocks. Where did the mixture come from? How did the rock change?

To unravel the story of this rock, we must return to the ocean floor. We must look closely at the area where the ocean floor collides with the continent. All of the various rocks that are transported, carried, lifted, or dropped at the edge of the continent can become part of this mixture. This whole jumble of rock and dissolved minerals is jammed into the trench at the top of the subduction zone. The rocks of the ocean floor can mix with the rocks from the continent in the trench.

If a sedimentary rock like conglomerate is caught in the trench and dragged below the continent, it can change. There is a lot of pressure on the rocks in the trench. The rocks can be reheated, pressed and welded together into one solid piece. Sedimentary rocks can become metamorphic rocks. These rebuilt chunks of rock can break off and be lifted by pressure from the rock layers sliding beneath them. You can be suspicious of this rock if it looks like a greenish conglomerate on the outside, but the inside pieces don't want to let go.

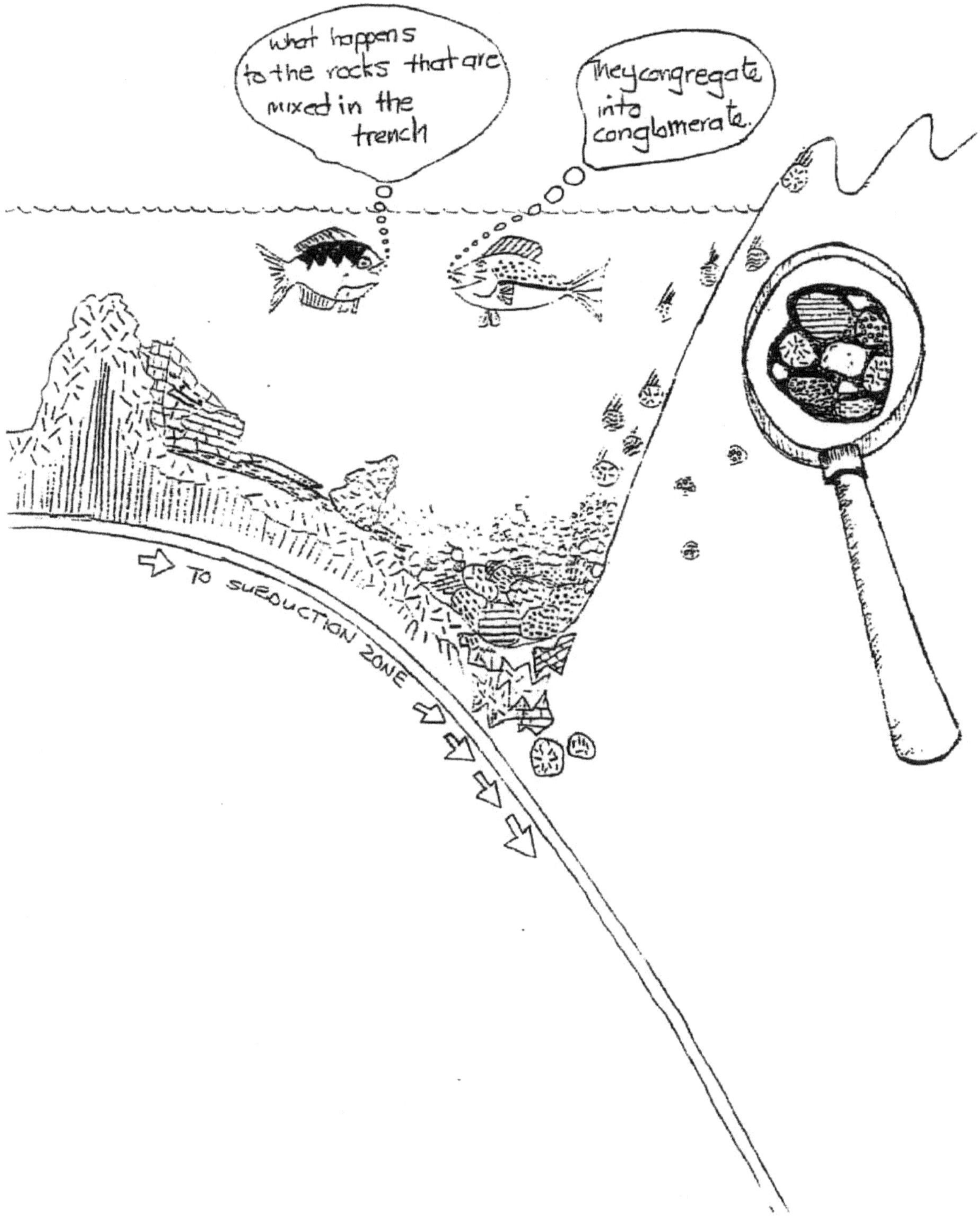
what happens to the rocks that are mixed in the trench
They congregate into conglomerate.
TO SUBDUCTION ZONE

Rain and rivers carry away many tons of rock. The hills are worn away. Thousands of earthquakes later, rock from the trench appears near the surface. The big story of the landslides and earthquakes or pressure in the trench is locked inside the rock. Each separate small rock that makes up a conglomerate has its own smaller story to tell. The small rock inside may be a piece of any rock that could find its way to the trench. If this were an ordinary sedimentary rock, we could break it apart into its original pieces. This greenish conglomerate is no ordinary sedimentary rock. It has been changed. It will no longer break into the original pieces. The shape of the pieces can be seen, but they will not break apart. The many pieces become one rock This rock is changed. It is now a metamorphic rock. Close inspection could show that new minerals are in the rock. The chemical changes can help explain the change in color and hardness.

Sometimes, the pieces in the rock are not rounded at all. The rock is made from smaller rocks with corners and edges all welded together. These rocks are known as breccia. Broken pieces make up breccia. These pieces did not have time to roll around and smooth out their edges like conglomerate. If you look closely you can figure out that part of their life story also.

SANDSTONE

If you wanted to make a sedimentary rock, one of the easiest ones to make would be sandstone. It would be easy to make because the pieces are easy to find. Sand is a common size of sediment. You can find it at beaches. The particular type of sandstone that is found around old subduction zones is called greywacke. It is made of many dark brown or grey pieces of large and small sand. Volcanoes made from dark rock, like basalt, are common around active subduction zones. Weathering and erosion wore away the volcanoes and brought the dark pieces to the beach.

Most of the common rocks that form on the land are made from sediment like sand or clay. Layers of rocks, gravel, sand, and clay pile up year after year on the edges of the continent. Rivers bring rocks, gravel, sand, and clay down to the ocean. The rocks may hold secrets about sand dunes, rivers, marshes, or great underwater landslides. These pieces of rock and sediment can be carried by currents and rolled farther out to sea along underwater canyons. The sediment on the bottom of the pile will eventually turn to rock.

It is not unusual to have violent collisions between the moving block of ocean floor and the continent. Strong earthquakes cause landslides on steep slopes on the land and underwater. There is a special event that happens underwater at the edge of the continent. When great heaps of sand, clay, rock and water slip, they become an underwater landslide officially known as a turbidity flow or density current. They can spread out over a hundred miles. Since this muddy water is more dense than sea water, it sinks and carries everything to the ocean floor dropping the largest rocks first. Sand sized pieces fall later. The last pieces to fall are the smallest. This makes a nifty pile of sediment sorted into layers by size. Rocks formed from density currents of turbidity flows are turbidites. They are records of underwater landslides. Sandstone is a common rock in these piles.

BLUESCHIST
(a metamorphic rock)

This rock is commonly a greyish sparkly-looking rock, but on the inside, it is sometimes bright blue It can have many flat, shiny minerals in it. One of these flat, shiny minerals is **mica.** Mica is easy to recognize. It is thin, shiny, flexible and only about as hard as a fingernail. Most rocks can scratch mica.

Blueschist kept its life story a secret for a long time. Some of it started in the ocean as basalt. It spent some time as Greenstone traveling along with the ocean floor. This chunk of floor continued its journey toward the edge of the continent, where it bent and slid into the trench. Unlike some of the rocks above it, this bit of rock did not get scraped off. This rock that would become blueschist, slid below the continent. The rock went deep into the subduction zone. It went ten or twenty miles below the surface. The pressure from all the rocks pressing down from above was so great that the minerals in the original stone changed position, shape and composition. The pressure and heat in the subduction zone are enough to make new minerals out of old ones. This happens to all the rocks that are caught in the subduction zone. Deep in the subduction zone, rocks turn to blueschist. Most of it keeps sliding into the hot earth until it is completely melted, and nobody ever sees it. Once in a while, a chunk breaks off. It stops sinking. Other rocks slide below it and begin to push it up. Someday this rock with the dramatic life story appears on the surface. It tells a story of how to get into the subduction zone and survive.

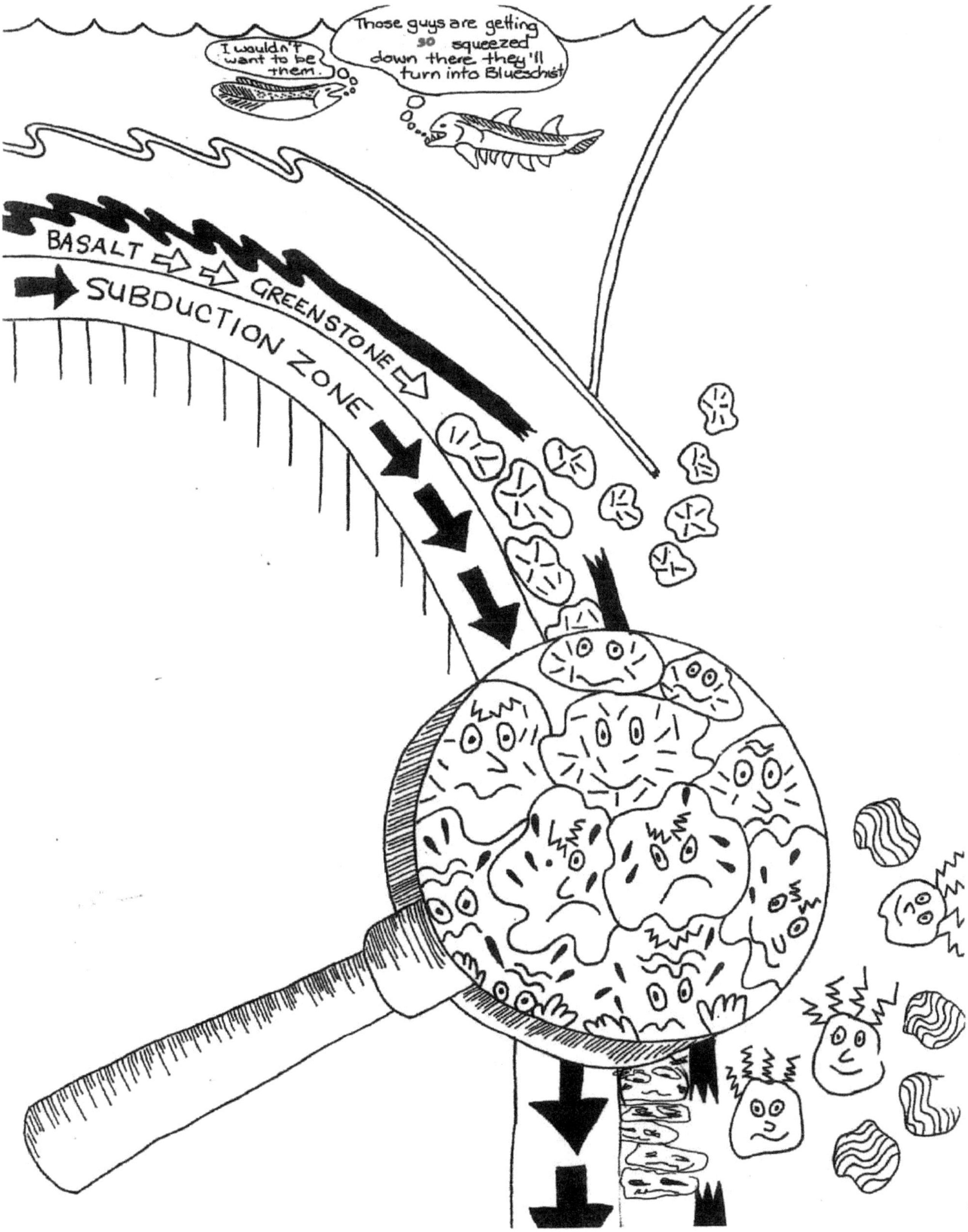

I wouldn't want to be them.
Those guys are getting so squeezed down there they'll turn into Blueschist
BASALT
GREENSTONE
SUBDUCTION ZONE

Some rocks do not give up their secrets easily. They have to be coaxed with special equipment. When scientists were trying to figure out the story of blueshist, they built a special machine to squeeze rocks.

They put a small piece of basalt in the machine. They added more pressure. At last, new minerals began to appear. With enough pressure, new minerals, such as glaucophane, began to grow. Glaucophane can only form in the special conditions found in a subduction zone where pressure is high, but temperature is relatively low. Because rocks do not heat up quickly, it takes a while for the cold ocean floor to become hot as it dives beneath the continent.

Now, scientists have more evidence of the great pressures that exist in the subduction zone. The life story of the blueschist was no longer a secret. Geologists who specialize in these things can tell how deep or how hot it was in the subduction zone where the changes took place.

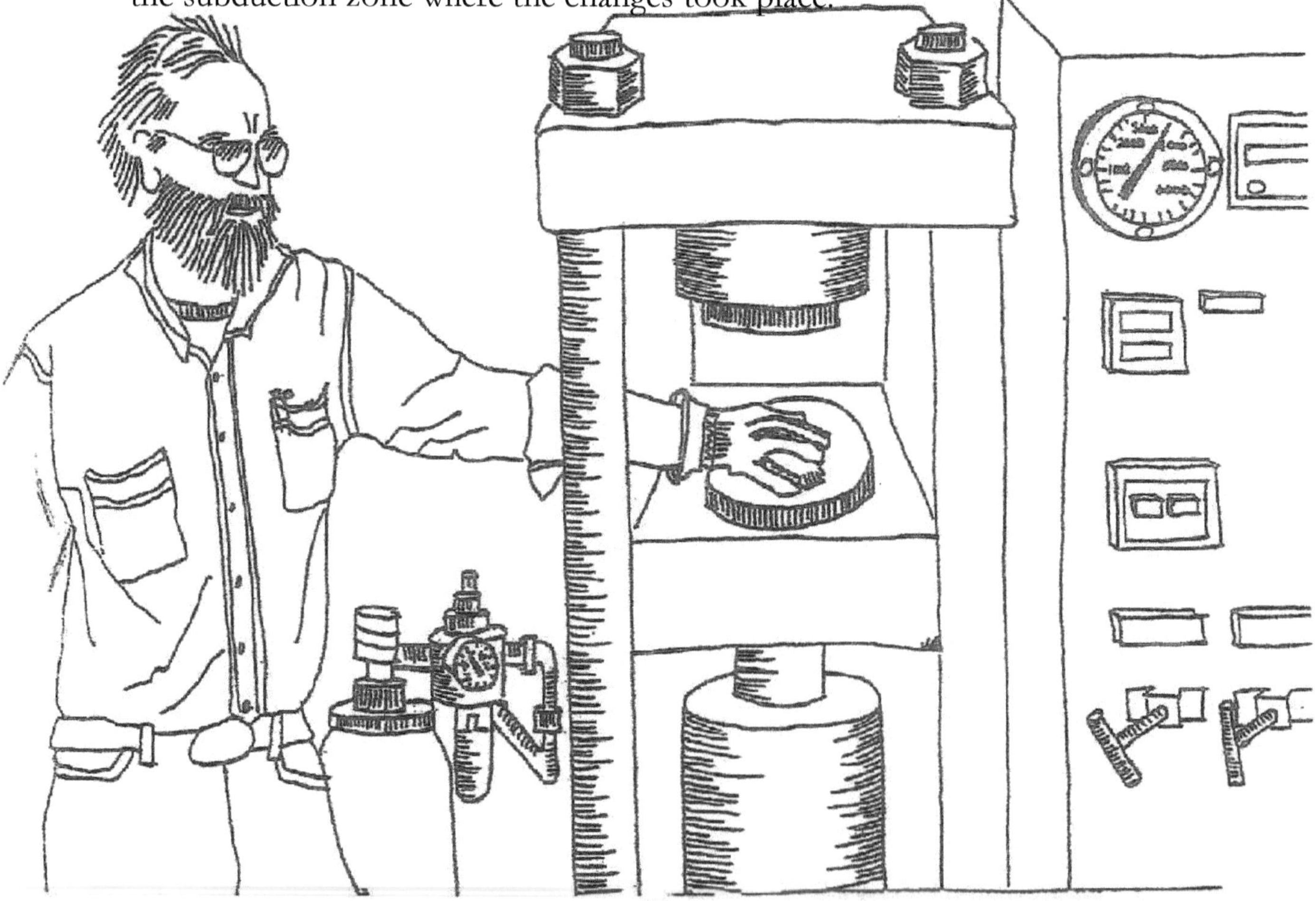

SERPENTINITE (a metamorphic rock)

This rock happens to be the official rock of the state of California. Pioneers crossing the country did not see rocks like this on the prairie or in the Rocky Mountains. The rock was named long before anyone understood where it came from. Serpentinite is named for the serpents that the people thought of when they saw wavy lines in the green slippery rock. The San Francisco end of the Golden Gate Bridge is anchored in serpentinite. Why do you find pieces of serpentinite in places like California, the Aleutian islands or Staten Island, New York?

Serpentinite forms when water is added to the dark, dense rocks found in Earth's mantle, the layer that lies below the ocean floor. It took big collisions of the sea floor and the land to break off chunks of this part of the ocean floor. Serpentinite is never the first thing scraped off. Even if you could get to the ocean floor alive, you still wouldn't see this rock. You would need to drill a hole with a special tool. Drill through the chert. Drill through the basalt and greenstone. Keep drilling. Now, look down towards the bottom of the hole. You would see the green crystalline minerals that makeup serpentinite.

These deep minerals can change when they have heat from below and water from above. Ocean water can follow cracks in the rocks and reach down to the minerals. The heat and water make it possible for the original rock to change to serpentinite. No one could tell the whole story of this rock until they could explain how rocks from the bottom of the ocean floor could get up into the hills.

WHEN HEAT AND
WATER ARE JUST RIGHT
DEEP ROCK WILL FORM
SERPENTINITE
THEY TELL
ME THAT ITS
GREEN
WHEN SEEN

THE FRANCISCAN ASSEMBLAGE

The slow motion of the ocean floor caused constant pressure at the boundary between the ocean floor and the continent. The pressure causes the smooth layers of rock that form on the ocean floor to bend and break and slip and twist. Rocks that form in the ocean are mixed with rocks that form on land. Geologists could have called these rocks the subduction zone jumble. The whole group of rocks along the California coast that includes rocks from the ocean floor and rocks from the land is called the Franciscan Assemblage. The rocks are named after San Francisco, the first place the rocks were found to be very common and recognized as a group. You can find these rocks all along the hills and valleys of the Coast Range in California.

The rocks in the hills along the coast have come from huge undersea lava flows, deep deposits of sediment, skeletons of plankton that have turned into solid rock, and pieces of the deep sea floor itself. These rocks were continuously carried to the subduction zone and mixed with pieces of the continent in the trenches. In this area of earthquakes and uplift, subduction and eruptions, weathering and erosion, the Coast Range was born.

The Franciscan Assemblage rocks tell the story of how the ocean floor was smashed against the edge of North America and created the land that would become California. The record of the old subduction zone is kept in the rocks of the coast range. Rocks like this are found in other places like Alaska, Japan, China and Canada. Such rocks can tell the story of other subduction zones, other times, and even other oceans.

The rocks and their disorderly arrangement were difficult for geologists to explain before the theory of **plate tectonics** was developed. Now, scientists have figured out the patterns of earthquakes and volcanic eruptions. The pieces of earth that move as a single unit have been identified. They are called plates. Tectonics means having to do with putting things together or building. Plate tectonics is the theory that moving and colliding plates have built hills and mountains and added pieces to continents. These pieces have assembled pieces of continents all over the world. The evidence is in the rocks.

Map of Current Plate Boundaries

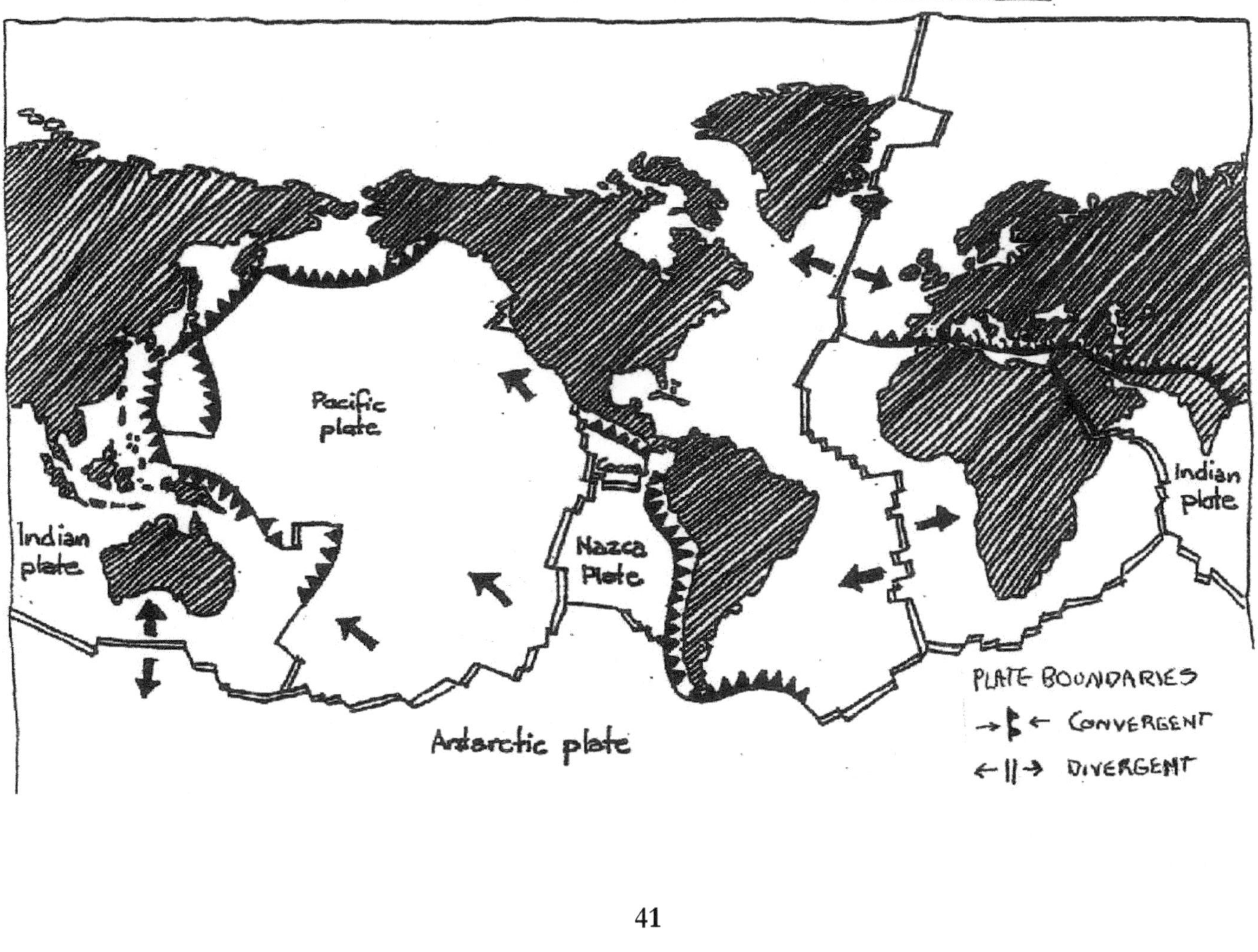

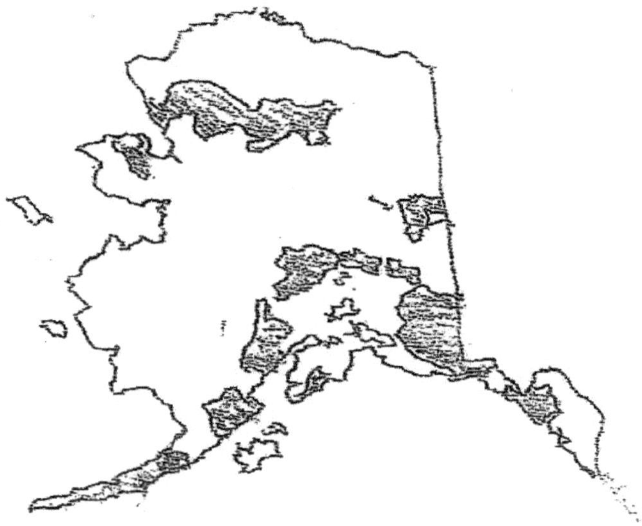

Other places with rock originally from the sea floor

Shaded areas contains sea floor rock from other boundaries

Map of places with subduction zone rocks

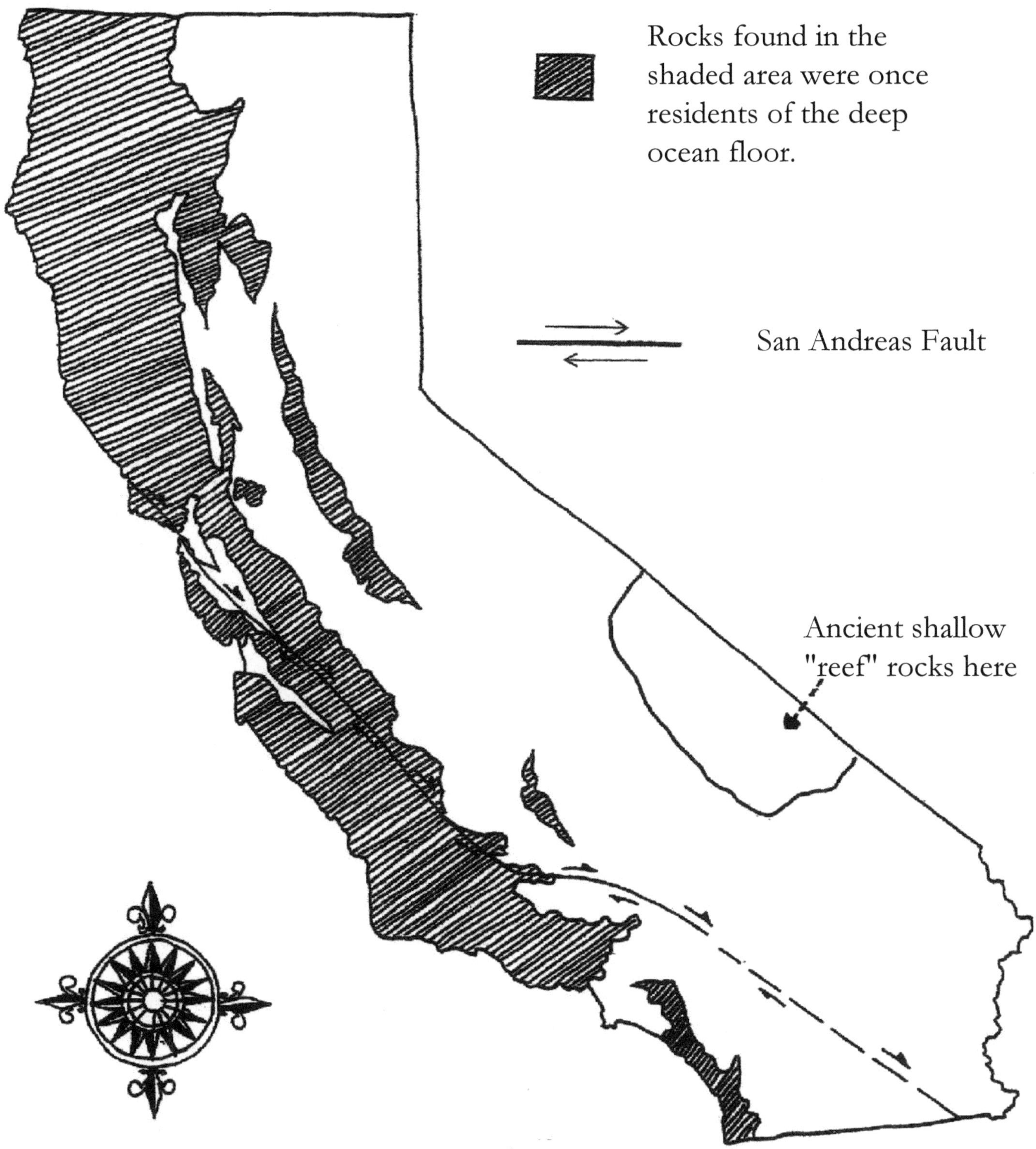

Today, the only subduction zone in California is way up north by Mendocino. The San Andreas fault and related faults form the boundary between the Pacific and North American plates. Rocks on the west side of the San Andreas fault slide north relative to the rocks on the east side, from Mexico all the way past San Francisco. This kind of boundary, where two tectonic plates slide past one another, is called a **transform boundary**. It is famous because it is a huge, long lateral fault. It moves sideways across the land. It doesn't duck underneath. People come from all over the world to study the work the San Andreas fault has done. Most of the boundary between California and the ocean is not a subduction zone. So, how could the rocks have come from the ocean? The rocks can explain it. The rocks keep very old secrets.

Do you know the secret? Go backward in time. Go back to the time before the San Andreas fault started to move. Now, Think about the boundary of the ocean and the continent. The whole Pacific plate was moving toward the east. Around 28 million years ago, the subduction zone began to change to a transform boundary. There is a little jog underwater in the line of The Hawaiian Islands as they head up to the Aleutian trench. This jog is like a bookmark in time. It tells you that the Pacific plate changed from a subduction zone to a lateral fault. The record of the old subduction zone is kept in the rocks of the California Coast Range

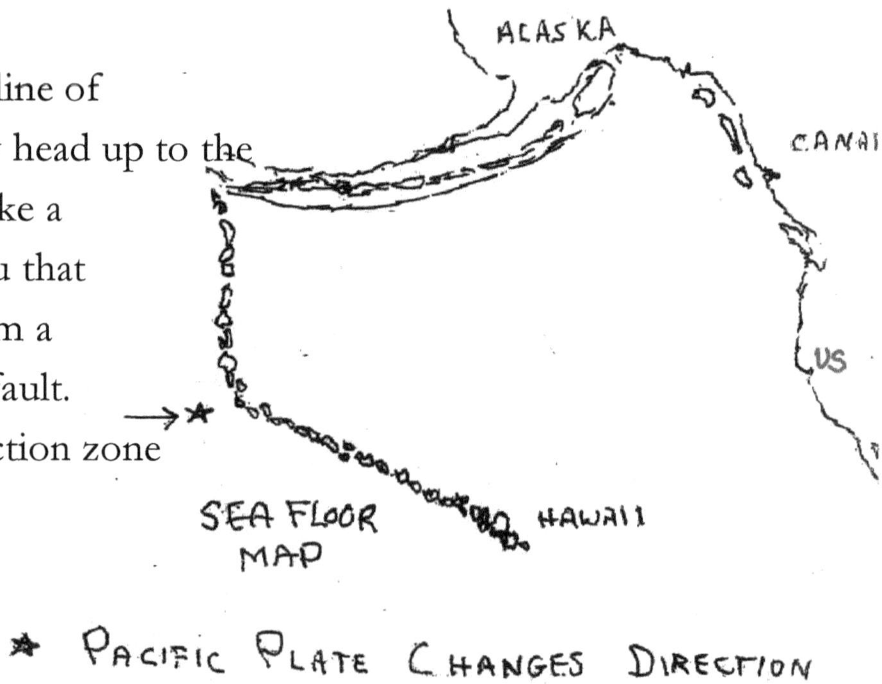

THE ROCK CYCLE

King Solomon was thinking of the water cycle about 3000 years ago when he said, "All the rivers run to the sea and to the place where the rivers began, there they return again". It took much longer to figure out the rock cycle. In the middle of the last century we were starting to put the evidence together.

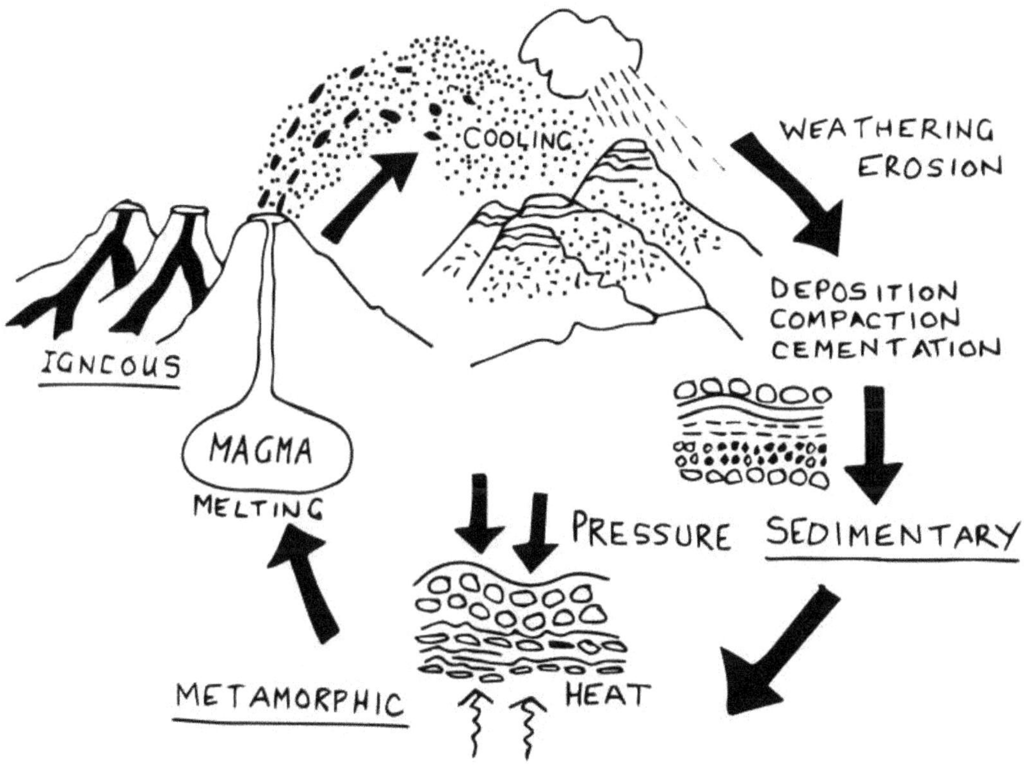

If a rock travels all around the rock cycle, it can tell all the stories in this book. There are also lots of shortcuts, just like in the water cycle. The secrets in the rocks don't evaporate. That is why we still have their stories.

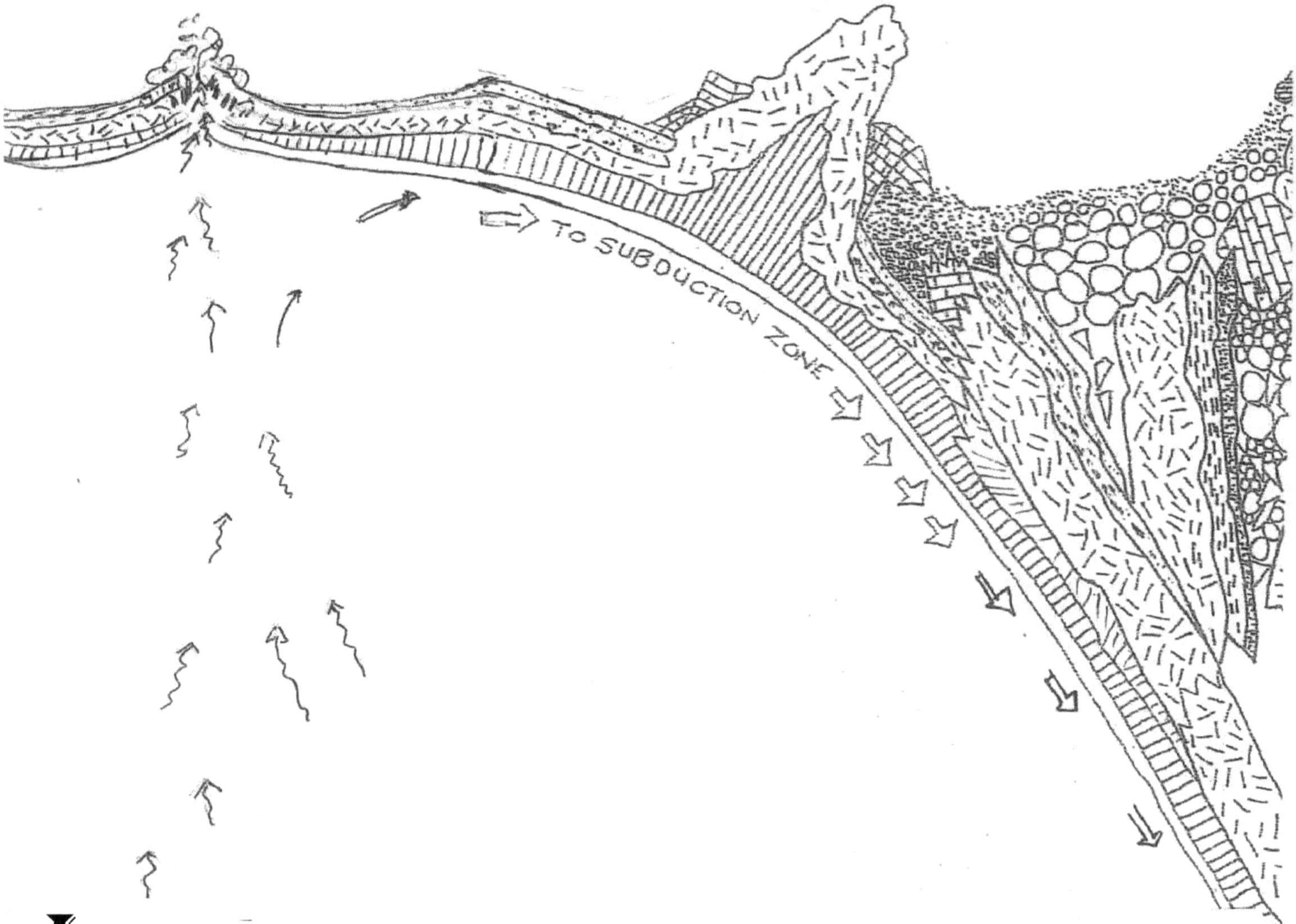

Look at the deepest part of a subduction zone. There, the melted rock has two choices. If the rock moves deeper and deeper, it will be caught in a huge convection current heading back to the spreading center. There it can start the cycle all over again.

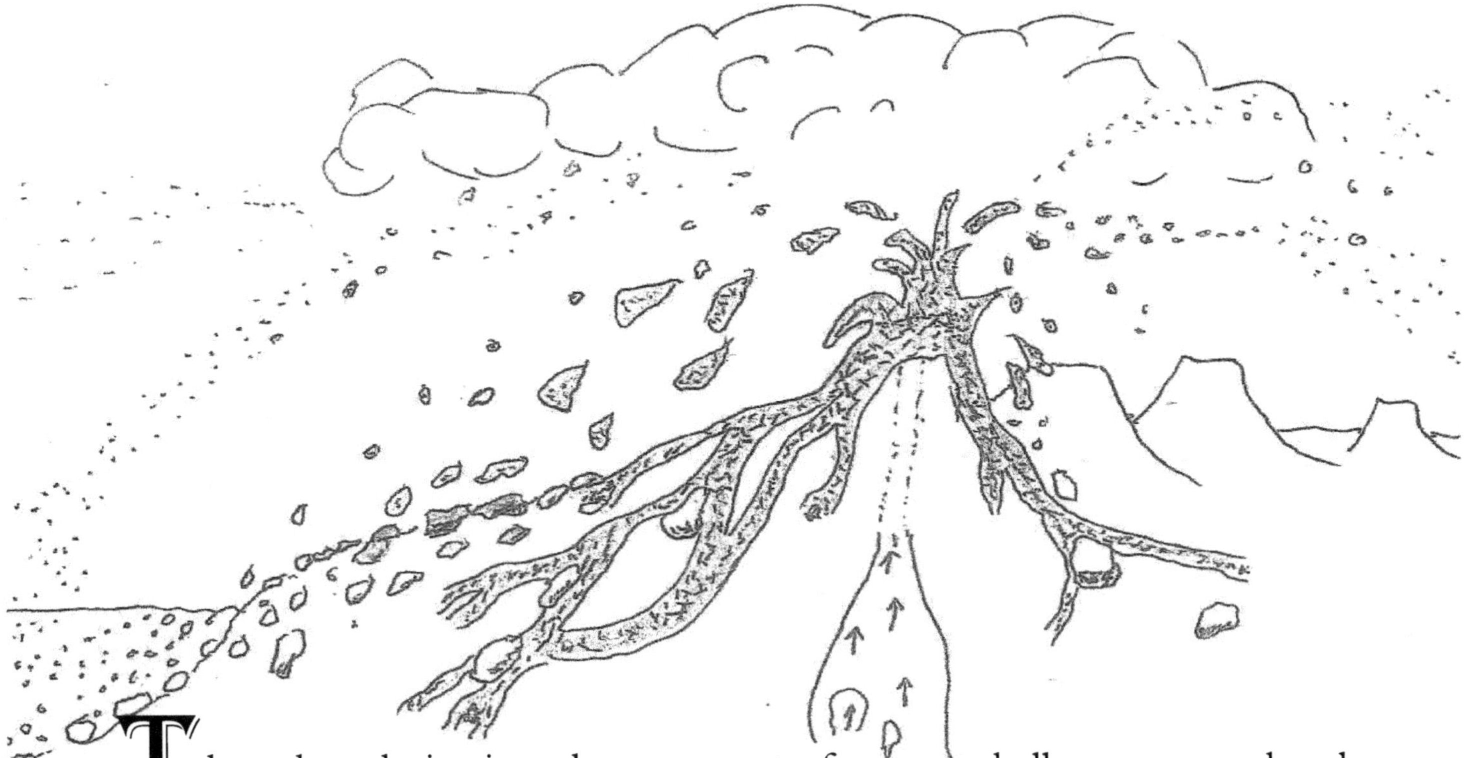

The other choice is to become part of a more shallow magma chamber. Minerals in this chamber are mixed around like something on the kitchen stove. The mixture can gain heat rising from below. The mixture can expand and lift up the land they are under. Just like a boiling mixture can lift the cover off the pot on the stove; so, a magma chamber can lift the cover off a volcano. Lava spills out or explodes. Heat and pressure are released. There you have a volcanic eruption. The pieces from volcanic eruptions, with the help of wind, water and gravity, can even find their way to other countries and oceans. There are still more secrets.

This is an exciting part. When the magma chamber is expanding as it lifts the mountain, cracks can form. What do you suppose fills in the cracks? Amazing and true. The minerals that are hardest to melt, cool first. Minerals that are easier to melt stay melted and rise to the top, like melted chocolate. These melted minerals are pushed around with minerals like themselves. Sometimes, when these minerals with low melting points rise to the top, they are pushed into cracks and fill in the empty spaces. Eventually, it all cools off. When miners find these filled spaces, also known as **veins**, they get very excited.

Some of the former cracks are filled with copper. That's how Copperopolis got its name. The whole area was filled with veins of copper. Weathering and erosion can uncover these veins. As the mountain wears away, rain can carry pieces of quartz, copper and gold downstream. That is how the California gold rush got started. First, miners found the gold in the river. Then, they followed the evidence upstream. Another valuable mineral in these veins is silver. The Comstock Lode in Nevada tells a story about Silver and the Wild West.

The rocks had the story long ago. Rocks have more secrets to tell, just waiting for someone to ask the right questions

Common Rock Symbols

SEDIMENTARY ROCKS

breccia

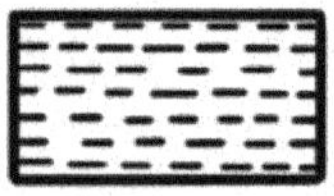

chert

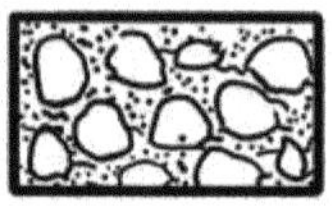

conglomerate

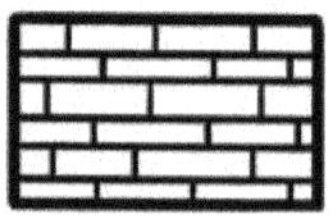

limestone

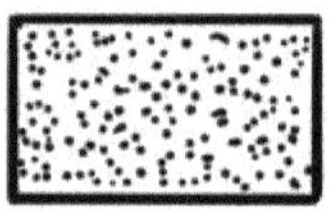

sandstone

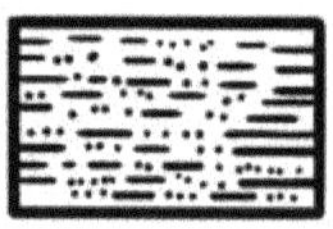

shale

IGNEOUS ROCKS

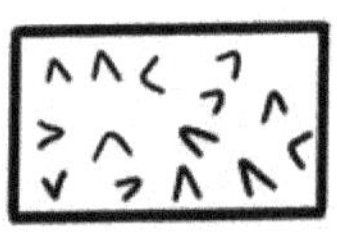

basalt

basalt columns

METAMORPHIC
ROCKS

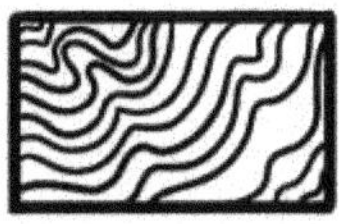 blueschist

 greenstone

 serpentinite

GLOSSARY

BASALT a dark igneous rock, a basic building block of the ocean floor.

BRECCIA (brech'shia) a rock made of broken angular pieces of other rocks all cemented together.

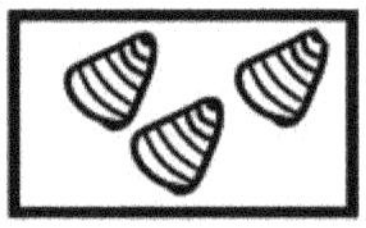

CALCITE a common rock forming mineral,a common material in skeletons of many sea creatures like coral and clams.

CRYSTAL the shape of the minerals when it has time and space to grow.

EXOSKELETON a skeleton found on the outside of creatures like coral and snails.

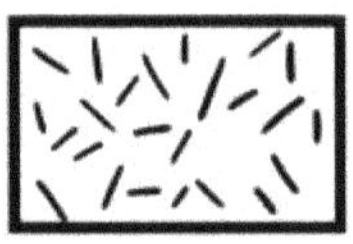

IGNEOUS ROCKS a rock that forms form cooling magma.

LAVA melted rock after it flows out onto the Earth.

MAGMA hot melted rock formed inside the Earth

MANTLE a hot layer of material below the crust of the Earth

METAMORPHIC ROCKS a rock that changes from igneous,sedimentary, or other metamorphic rocks as a result of changes in pressure or temperature.

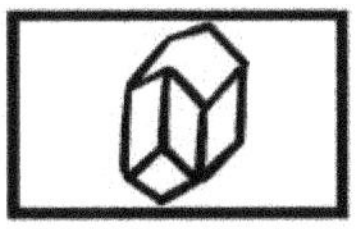

MINERAL a naturally accurring element or compound like calcite or feldspar, the small units from which rocks are made

OOZE the gooey sediment that collects on the oceans floor, mostly clay and skeletons of radioloarians and other planktons

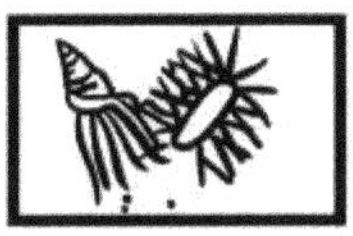

PLANKTON microscopic creatures that float and drift in the top layers of the ocean

PLATE TECTONICS the theory that sections of the earth move as units and build new land forms along the boundaries

RADIOLARIAN a type of plankton with a silica based skeleton

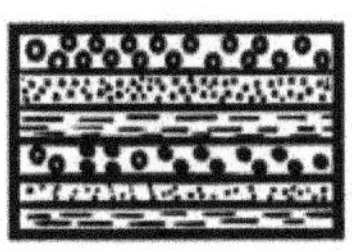

SEDIMENTARY ROCK a type of rock formed from pieces of sediment like clay and sand, commonly found in layers

SILICA a mineral common to many rocks chert and some plankton like radiolarians

SILT small muddy pieces of sediment larger than clay just barely big enough to be seen without a microscope.

SPREADING CENTER the area where new ocean floor is pushed up. Older floor moves away from the spreading center.

SUBDUCTION ZONE (sub duck shun) the area where the ocean floor sinks down into the soft hot mantle of the earth beneath the edge of the continent.

TRANSFORM BOUNDARY a boundary where two tectonic plates slide past each other.
Example: North American plate and pacific plate

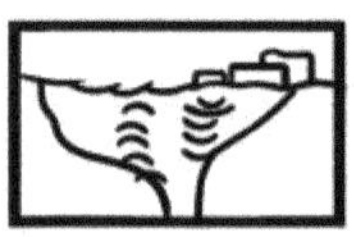

TRENCH a deep, narrow depression in the ocean floor caused by the sinking of the ocean floor below the continent.

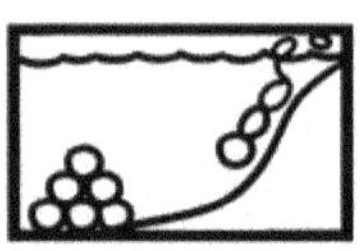

TURBIDITIES rock formations made of sediments that have been sorted by size during underwater landslides. The largest pieces are on the bottom. The smallest pieces are on the top

VEINS cracks where minerals like gold or silver have collected

WEATHERING changes that take place on the surface that weaken the rocks like rusting or freezing and cracking.

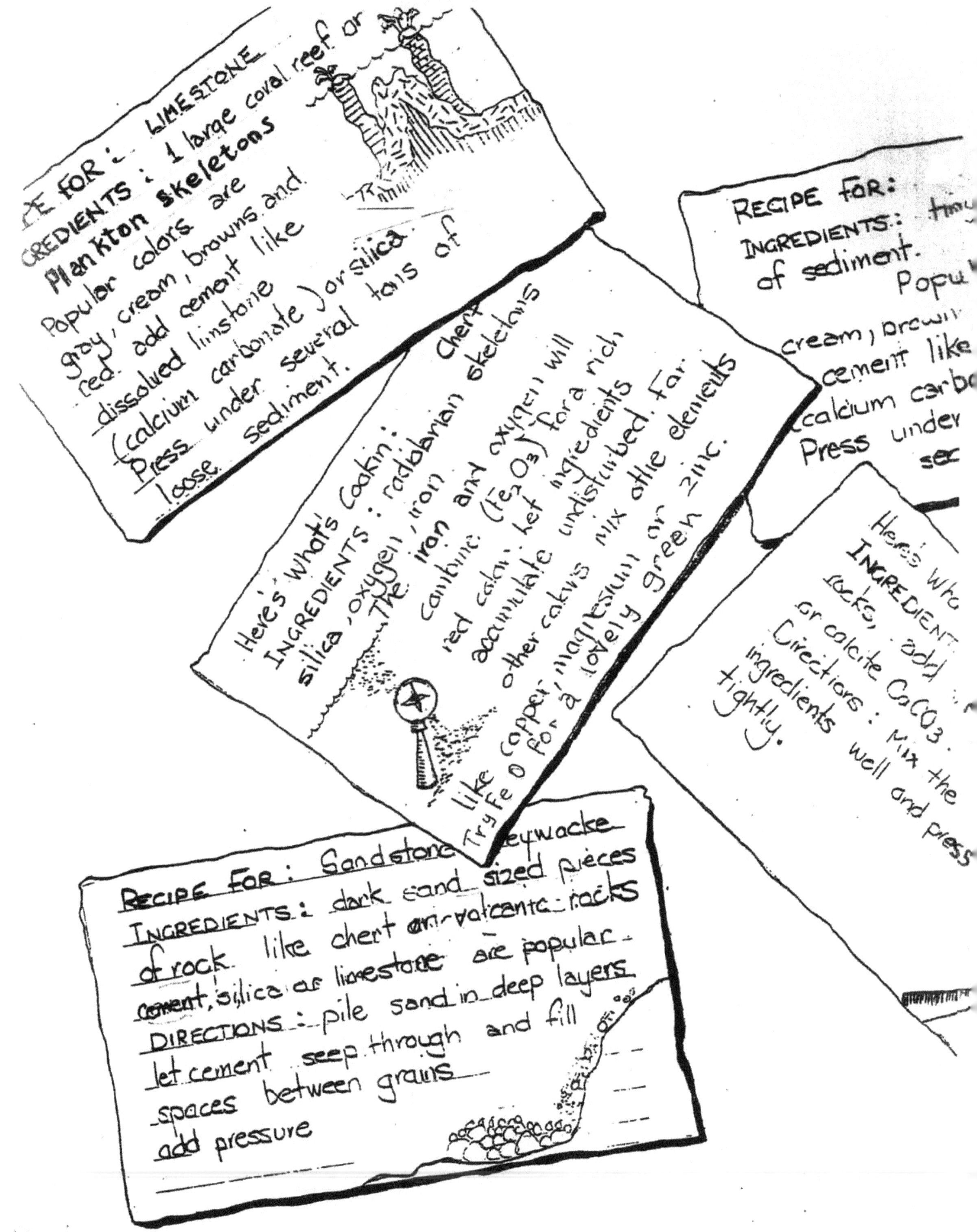
PE FOR: LIMESTONE
GREDIENTS: 1 large coral reef or
Plankton skeletons
Popular colors are
gray, cream, browns and
red. add cement like
dissolved linstone
(calcium carbonate) or silica
Press under several tons of
loose sediment.
Here's What's Cookin: Chert
INGREDIENTS: radiolarian skeletons
silica, oxygen, iron
The iron and oxygen will
combine (Fe2O3) for a rich
red color. Let ingredients
accumulate undisturbed. For
other colors mix other elements
like copper, magnesium or zinc.
Try FeO for a lovely green
RECIPE FOR:
INGREDIENTS:
of sediment.
cement like
(calcium
Press under
Here's
INGREDIENT
or calcite CaCO3.
Directions: Mix the
ingredients well and press
tightly.
RECIPE FOR: Sandstone greywacke
INGREDIENTS: dark sand sized pieces
of rock like chert or volcanic rocks
are popular
cement, silica or limestone
DIRECTIONS: pile sand in deep layers
let cement seep through and fill
spaces between grains
add pressure

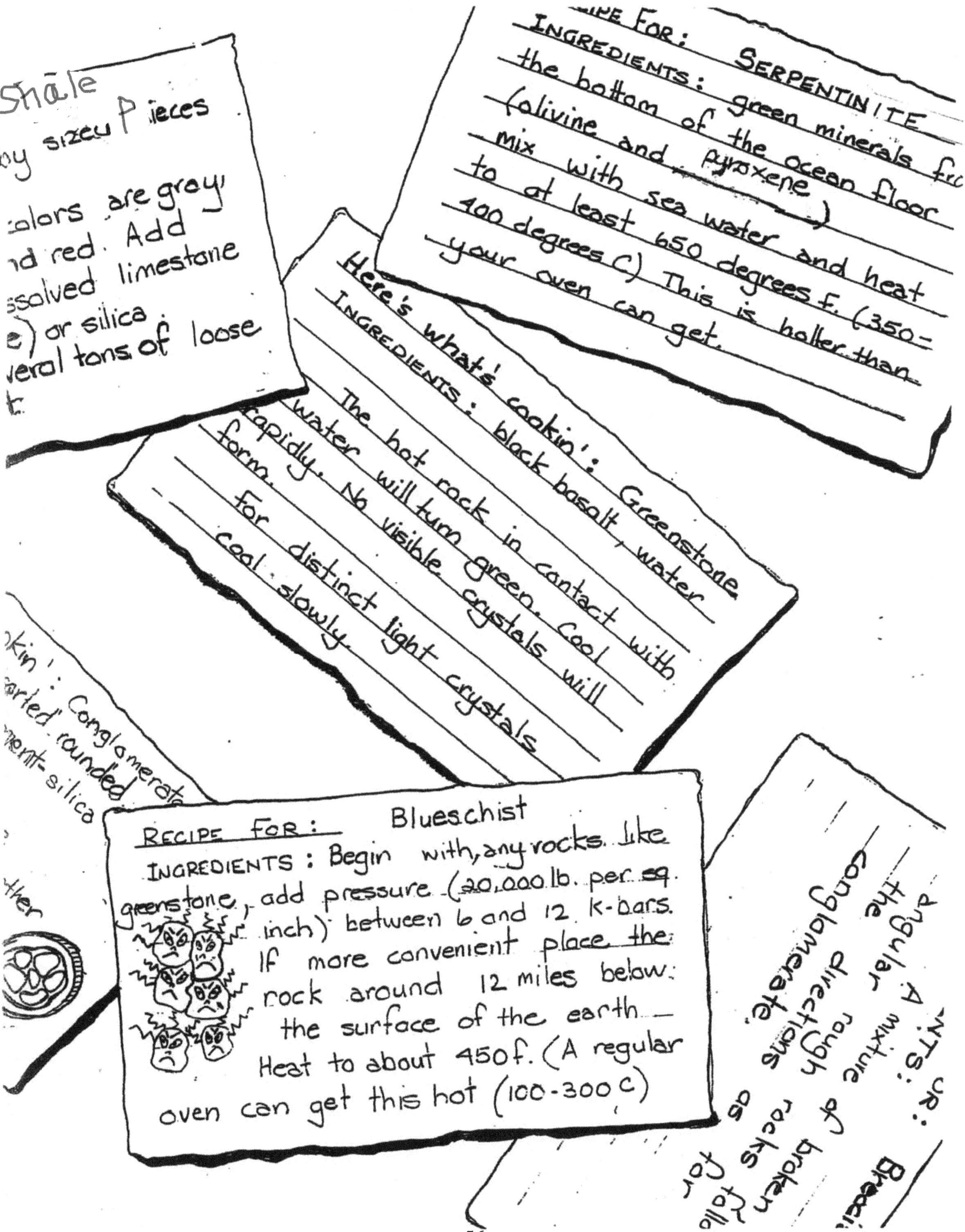
Shale
Pieces
colors are gray
nd red. Add
ssolved limestone
e) or silica
veral tons of loose
INGREDIENTS: SERPENTINITE
green minerals from
the bottom of the ocean floor
(olivine and pyroxene)
mix with sea water and heat
to at least 650 degrees F. (350-
400 degrees C) This is hotter than
your oven can get.
Here's what's cookin': Greenstone
INGREDIENTS: black basalt, water
The hot rock in contact with
water will turn green. Cool
rapidly. No visible crystals will
form.
for distinct light crystals
cool slowly.
kin': Conglomerate
rted" rounded
ment-silica
RECIPE FOR: Blueschist
INGREDIENTS: Begin with any rocks like
greenstone, add pressure (20,000 lb. per sq.
inch) between 6 and 12 k-bars.
If more convenient place the
rock around 12 miles below
the surface of the earth
Heat to about 450f. (A regular
oven can get this hot (100-300 C)
Breccia
NTS: A mixture of broken
angular rough rocks
the directions as
conglomerate.

AUTHOR BIO

Dianne Connelly MA, fifty years of teaching experience, high school and Jr. High, mostly science and math. The book I have written was the result of a federal grant to make our high school a science and math magnet as an alternate to busing. The book, Sea Floor Stories, began as field trip notes. I couldn't find a book that would explain the presence of the mixture of rocks in the Sierra foothills and coast range. I have had help writing from San Jose State, USGS, and many others listed in acknowledgments.

If you could go back in time about 360 million years you would see the Pacific Ocean on the beach of Nevada. California was being brought to North America slowly, compliments of the sea floor. The rocks on the west coast of North America have a wonderful story to tell

AFTERWORD

During World War ll, when ships and submarines were being sunk, the United States among others gathered some interesting data about the sea floor. The information was classified so that no one could use it to sink more ships. This page is in the afterward of the book because I couldn't find some of some of the information I wanted until after it was declassified in the 90's.

Lyn exploring a sandstone outcrop.
Madeline Island, Wisconsin

John and Andrew scrambling on layers of shale and chert.
Point Reyes, California

ILLUSTRATOR BIO

Lyn Reed MA, has over 40 years of teaching experience in Art Education. She was born in Australia, and now works as a printmaker and ceramic artist in Northern Minnesota.

Lyn met Dianne Connelly in Australia where they taught in the same school and became life long friends. When Dianne needed someone to illustrate Sea Floor Stories, it seemed a fun and natural collaboration..

Lyn is an avid rock hound herself-traveling throughout Australia and the U.S with her husband, Mark, and sons John and Andrew, collecting rocks along the way.

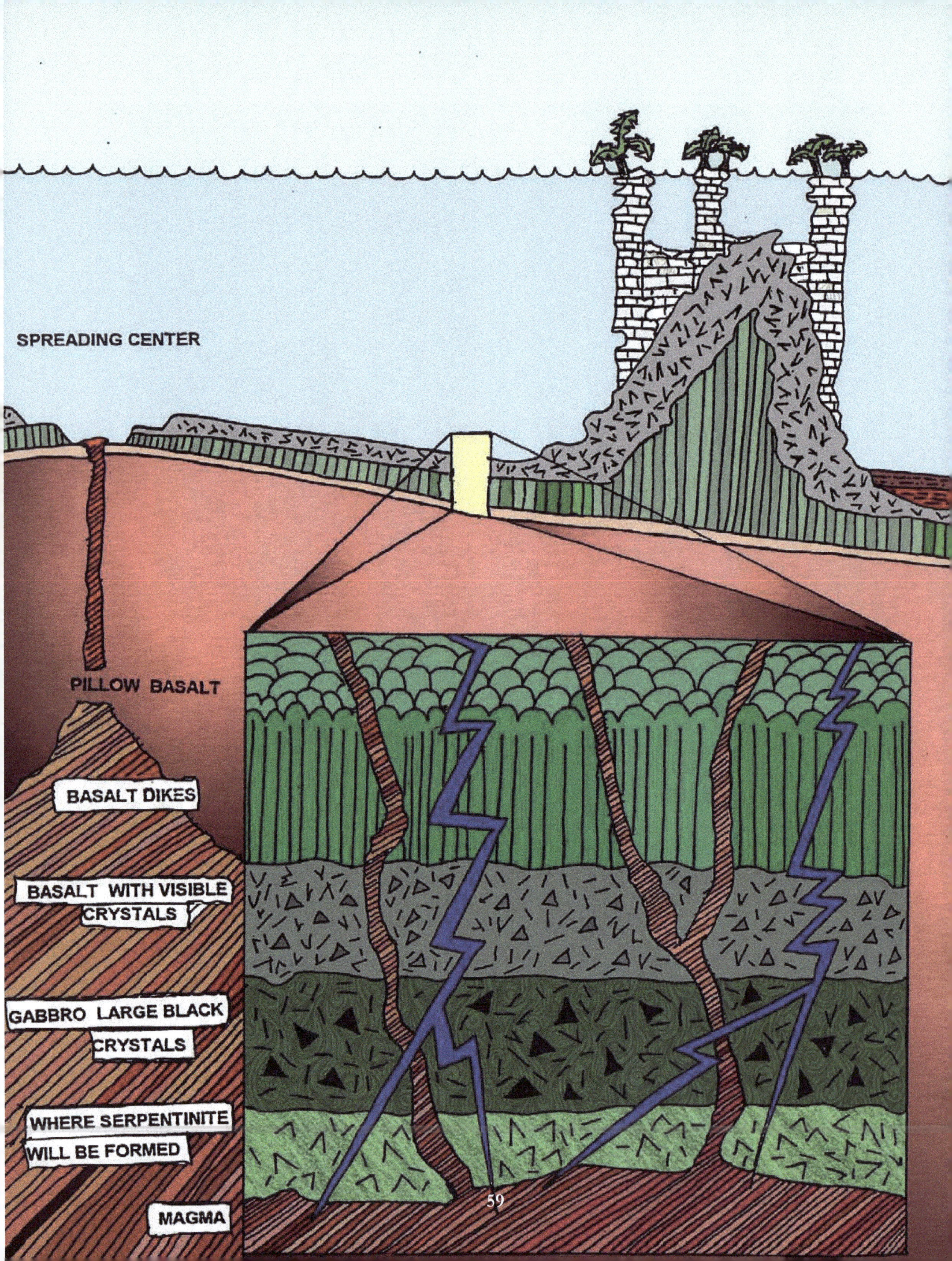
SPREADING CENTER
PILLOW BASALT
BASALT DIKES
BASALT WITH VISIBLE CRYSTALS
GABBRO LARGE BLACK CRYSTALS
WHERE SERPENTINITE WILL BE FORMED
MAGMA

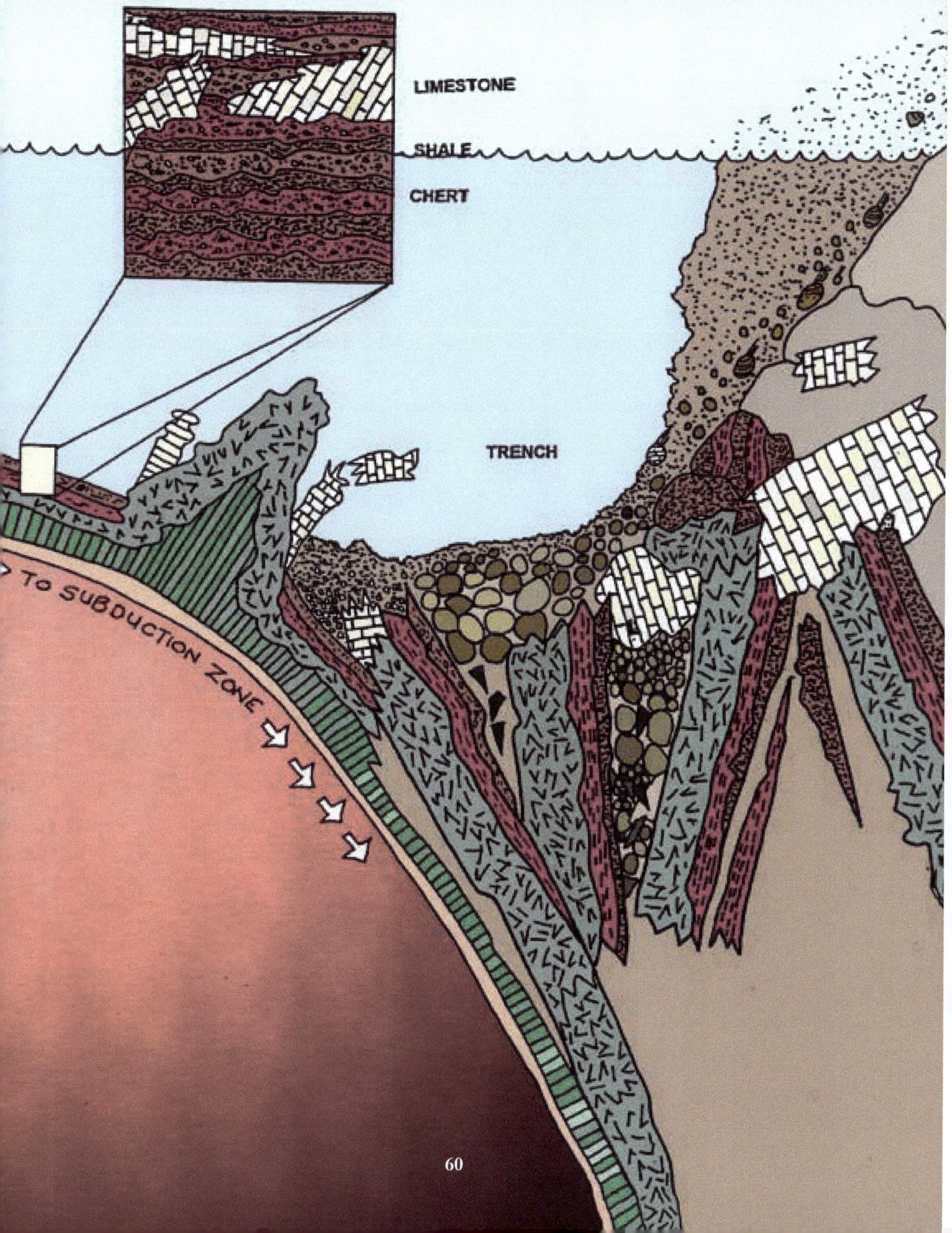
LIMESTONE
SHALE
CHERT
TRENCH
TO SUBDUCTION ZONE

ACKNOWLEDGMENTS

Without the assistance and resources listed below, I would not have been able to write this book. I would like to thank the United States Geological Survey office in Menlo Park for their resources and help, particularly Ken Lajoie, for his encouragement, Phil Stoffer for his recommendations and Ben Hankins for his help in the Magma dynamics lab.

Thanks to Stanford University for the current and accessible resources of the Ruth Mitchell Earth Science Library. I appreciate the assistance I received from the San Jose State Geology Department, particularly Dr. Ellen Metzger and Dr. Richard Sedlock.

Thanks to John Molstad, Alan Stephenson, Robin McGinnity Connelly and many others for their ideas and suggestions. Thanks to John and Andrew Reed for their fish drawings and to the adult version of Andrew Reed as a geological consultant, Ali Jililian for his technical assistance; to the children who asked the questions that this book attempts to answer, and to my mom, who was always interested when I brought home a nice rock. Also, thanks to Lyn Reed, my illustrator, for her talent, patience, persistence and friendship. Thanks to many unnamed others whose interest and encouragement contributed to the eventual publication of this book.

www.ingramcontent.com/pod-product-compliance
Lightning Source LLC
LaVergne TN
LVHW070857210926
848482LV00016B/960

* 9 7 9 8 3 4 9 5 4 8 8 1 9 *